Praise for Alan Lightman's

PROBABLE IMPOSSIBILITIES

"Whatever the subject, [Lightman] writes with a limpid serenity and frankness that feel as fresh and as clarifying as a spring rain."
—Salon

"A series of essays on creation, consciousness, and our place in the universe, Lightman turns his attention to some of our biggest questions about infinity and nothingness."
—BuzzFeed

"A roaming, eye-opening, insightful, and literate collection of science writing.... Complex science made accessible."
—*Kirkus Reviews*

"Lightman's awe about the physical world is infectious. He speaks with the authority of a scientist, specifically a former Harvard astrophysicist, and the eloquence of a novelist.... *Probable Impossibilities* offers a primer on many of modern science's most mind-blowing discoveries, incorporating profiles of scientists performing cutting-edge research.... Lightman weaves his own story and voice through the book."
—*Publishers Weekly*

"Radiant. . . . Provocative. . . . Lightman, matching fact with awe, pilots readers on enlivening and enlightening thought voyages into such realms as quantum physics, genetics, miracles, and the expanding universe, each foray offering new coordinates, evocative vistas, and deepened understanding."
—*Booklist*

Alan Lightman

PROBABLE IMPOSSIBILITIES

Alan Lightman is the author of seven novels, including the international bestseller *Einstein's Dreams* and *The Diagnosis*, a finalist for the National Book Award. He has taught at Harvard and at MIT, where he was the first person to receive a dual faculty appointment in science and the humanities. He is currently professor of the practice of the humanities at MIT.

Also by Alan Lightman

PROBABLE IMPOSSIBILITIES

Musings on
Beginnings
and
Endings

Alan Lightman

Vintage Books
A Division of Penguin Random House LLC
New York

FIRST VINTAGE BOOKS EDITION, APRIL 2022

Copyright © 2021 by Alan Lightman

All rights reserved. Published in the United States by Vintage Books,
a division of Penguin Random House LLC, New York, and distributed
in Canada by Penguin Random House Canada Limited, Toronto.
Originally published in hardcover in the United States by
Pantheon Books, a division of Penguin Random
House LLC, New York, in 2021.

Vintage and colophon are registered
trademarks of Penguin Random House LLC.

Portions of this book were originally published, in slightly different
form, in the following publications: "What Came Before the Big Bang?"
in *Harper's*, January 1, 2016 · "On Nothingness" appeared as "My Own
Personal Nothingness," in *Nautilus*, August 28, 2014 · "Atoms," in *Searching
for Stars on an Island in Maine* (New York: Pantheon, 2018) · "In Defense of
Disorder," in *Aeon*, April 15, 2019 · "Miracles" appeared as "Splitting the
Moon," in *Guernica*, September 15, 2015 · "The Anatomy of Attention,"
in *The New Yorker*, October 1, 2014 · "Smile," in *Science* 84, 1984 ·
"Immortality" appeared as "Consciousness," in *The Fabulist*, March 2018 ·
"Our Lonely Home in Nature," in *The New York Times*, May 3, 2014 · "The
Ghost House of My Childhood," in *The New York Times*,
August 22, 2015 · "Is Life Special?" appeared as "Is Life Special Just
Because It's Rare?" in *Nautilus*, October 15, 2015

The Library of Congress has cataloged the Pantheon edition as follows:
Names: Lightman, Alan P., author.
Title: Probable impossibilities :
musings on beginnings and endings / Alan Lightman.
Description: First edition. | New York : Pantheon Books, 2021. |
Includes bibliographical references (pages 185–197).
Identifiers: LCCN 2020013914 (print) | LCCN 2020013915 (ebook)
Subjects: LCSH: Cosmology. | Metaphysics. |
Philosophy of mind. | Physics—Philosophy.
Classification: LCC QB981 .L544 2021 (print) |
LCC QB981 (ebook) | DDC 523.1—dc23
LC record available at https://lccn.loc.gov/2020013914
LC ebook record available at https://lccn.loc.gov/2020013915

Vintage Books Trade Paperback ISBN: 978-0-593-08132-7
eBook ISBN: 978-1-5247-4902-6

Book design by Betty Lew

vintagebooks.com

Printed in the United States of America

Contents

PROBABLE
IMPOSSIBILITIES

Probable Impossibilities

I will tell you a thing that is both impossible and true. You were born from a tiny seed within your mother. And she was born from a tiny seed within her mother. And she from her mother. And so on, back and back through the dim hallways of time until we arrive at a particular cave in Africa, a hundred thousand years in the past, with a particular woman sitting by a fire. That woman knew nothing of cities or automobiles or electricity. But if we could follow her daughters through time, we would eventually arrive at you. If each of those daughters of daughters had pressed an inky thumb on a large piece of parchment, one following the other, there would today be several thousand thumbprints on that parchment, leading from that ancestral woman a hundred thousand years ago to your thumbprint today.

If this story does not seem impossible, or at least incomprehensible, let's go further back in time. According to modern analysis of the DNA of fossil animals, your ancestral mother descended from more primitive creatures, and those from more primitive, until we reach single-celled organisms squirming and gyrating in a primeval sea. And those first living organisms emerged from the billions of random collisions of lifeless molecules, by chance forming things that could spawn more of themselves and tap energy from the roiling sea. And before that, the ancient air of Earth—methane and ammonia and water vapor and nitrogen—blew over the seething volcanoes. And before, the gases swirled and condensed from a cloud in the primeval solar system.

I will tell one final story. Every atom in your body except for hydrogen and helium was made in stars long ago and blown into space when those stars exploded—much later to be tossed into the air and soil and oceans of Earth and eventually incorporated into your body. How do we know? Evidence supports the Big Bang theory, which holds that our universe began in a state of extremely high density and temperature and has been expanding and cooling since. In the first moments after $t = 0$, the universe was far too hot for atoms to hold together. During the first three minutes, the universe cooled enough for the simplest atomic nuclei, hydrogen and helium, to form, but was thinning out too rapidly to make carbon and oxygen and nitrogen and all the other atoms our bodies are made of. According to nuclear physicists, the formation of those atoms occurred hundreds of millions of years later, when gravity was able to pull together large masses of gas to form stars. The temperatures and densities at the centers of those masses again began to mount, starting nuclear reactions, which fused the existing hydrogen and helium atoms into the other atoms in our bodies. Some of those stars exploded, seeding space with the newly forged atoms. With our telescopes, we have seen exploding stars and analyzed the chemical composition of their debris. We have confirmed the theory. If you could tag all the atoms in your body and follow them backward in time, every atom, except for hydrogen and helium, would return to a star. We are as certain of this story as we are that the continents were once joined.

Less certain but supported by compelling calculations are the infinities, the infinity of the small and the infinity of the large. The unending world of ever smaller things within the atom, and the unending world of ever larger things, beyond our telescopes. Between these two endpoints of the imagination are we human beings, fragile and brief, clutching our thin slice of reality.

Between Nothingness and Infinity

In a lifetime, most people travel no farther than five hundred miles from home. During that limited exploration of the physical world, we record memories of nearby objects and experiences—people, houses, trees, local lakes and rivers, sounds of birds, clouds—all funneled into our brains by our eyes and ears. And yet think what we are able to *imagine*. Take, for example, Homer's epic tale of the voyage of Ulysses. At one point, Ulysses and his men are captured by the Cyclops, a thirty-foot-tall man with a single eye in the middle of his forehead, who immediately eats two of the crew and imprisons the rest in his cave for future meals. On the seas again after his escape, Ulysses ties himself to the ship's mast so that he can resist the call of the Sirens, creatures with the bodies of birds and the heads of women, whose beautiful song lures men to their doom on the reefs. Or consider Salvador Dalí's famous painting *The Persistence of Memory*, in which rubbery clocks droop over tree branches and tables like pizzas melting in the

Sun. Horses with wings, rivers of gold, wooden puppets that come to life. In our minds, we have the power to combine things that we've seen in our paltry experience to create spectacular apparitions never before encountered, and even things that do not exist.

Imagination in the arts is familiar. Imagination in science not so familiar, yet breathtaking in its daring and frequent validation. Following the logical trail of his equations, James Clerk Maxwell imagined waves of electromagnetic energy traveling through space—X-rays and radio waves invisible to the eye. Einstein imagined that clocks moving past each other would tick at different rates, although such a preposterous phenomenon had never been observed. (To measure the effect, highly sensitive instruments are required, or relative speeds approaching the speed of light.)

The ancient Greeks hypothesized invisible atoms, minuscule things too small to see, indestructible, indivisible, the presumed building blocks of the material world—another leap of the imagination. Two thousand years later, a Frenchman named Blaise Pascal (1623–1662) imagined much further. Mathematician, physicist, inventor, essayist, and theologian, Pascal conjectured the existence of things *infinitely* small and *infinitely* large. From his *Pensées:*

> The whole visible world is only an imperceptible atom
> in the ample bosom of nature...We may enlarge our
> conceptions beyond all imaginable space; we only pro-
> duce atoms in comparison to the reality of things. It is
> an infinite sphere, the center of which is everywhere, the
> circumference nowhere...What is man in the infinite?

But to show him another prodigy equally astonishing, let him examine the most delicate things that he knows. Let a mite be given him, with its minute body parts and parts incomparably more minute. Dividing these last things again, let him exhaust his powers of conception and let the last object at which he can arrive be now that of our discourse. Perhaps he will think that here is the smallest point of nature. I will let him see therein a new abyss... For who will not be astounded at the fact that our body, which a little while ago was imperceptible in the universe... is now a colossus, a world, or rather a whole in respect of the nothingness which we cannot reach? He who regards himself in this light will be afraid of himself, and observing himself sustained in the body given him by nature between these two abysses of the Infinite and Nothing will tremble at the sight of these marvels... [Man] is equally incapable of seeing the Nothing from which he was made, and the Infinite in which he is swallowed up.

At the time that Pascal wrote this remarkable passage, the first crude microscopes had only recently been invented, and the greatest distance measured was that to the Sun. In particular, the size of the crystalline "heavenly sphere," on which hung the stars, was completely unknown. Working in a cold and dimly lit house on the outskirts of Paris, at a time when forced bleeding was thought to cure illness and the medicine cabinet brimmed with mercury and arsenic, when fire and electricity were totally mysterious, Pascal imagined infinity.

It was not only the physical infinities that captured Pascal's imagination. He was also concerned with the manner in which we human beings situate ourselves in the world, trapped in the bodies given us by nature, "caught between the two abysses of the Infinite and Nothing." One finds no such human consider-ations and poetry in the writings of Newton several decades later. Pascal was practically unique in being a humanist and a scientist at once—a close observer of human nature in such works as "The Misery of Man Without God" (Pascal was deeply religious), a worldly man born into the upper levels of French society, and a guest of the salons of Paris—while at the same time a mathematician who made seminal contribu-tions to projective geometry, an inventor who designed one of the first mechanical computers, a pioneer in the theory of probability. A unit of pressure, the pascal, is named after him. As is a computer programming language. One might compare Pascal to another great polymath of the Renaissance, Leo-nardo da Vinci. Yet even da Vinci did not contemplate infinity.

A famous portrait of Pascal, painted by his contemporary Philippe de Champaigne, shows a man of about thirty-five (Pascal died at age thirty-nine), with pink cheeks set against pale, sickly skin, hints of a delicate moustache and beard, a prominent and aristocratic nose, dark hair in curls hanging down to his shoulders, a green blouse like embroi-dered drapery slung across his chest, a starched white collar, and a vague, almost suffering smile, as if he is contemplating the misery of man without God and struggling to make the best of a sinful world.

Pascal was born into a well-to-do and devout family at

Clermont, in Auvergne. His father was a government official and tax collector. The young Blaise early showed precocity in mathematics and all things mechanical. While still a teenager, he started building calculating machines to help his father with his tax computations. After fifty prototypes, young Pascal succeeded in creating a finished machine, now called a "Pascal calculator." The gadget looks like a copper shoebox. Six windows show numbers, with six spoked metal dials below them. To input a digit, you place a stylus between the spokes of a dial and turn it until the number comes up in the corresponding window. Then you enter another digit in the next dial. By way of gears, the sum of the two numbers appears in another window.

At age sixteen, young Pascal taught himself geometry, helped by his drawing of charcoal figures on the flagstone floor. Soon, he discovered what is now known as "Pascal's theorem": If six arbitrary points are chosen on a conic section (the curve formed by the intersection of a plane with a cone) and joined by line segments in any order to form a hexagon, then the three pairs of opposite sides of the hexagon meet at three points (G, H, and K in the figure below), which lie on a straight line. I am not aware of any practical applications of Pascal's theorem, although it has been advertised as a great theorem to prove for the world's brightest high school students in the International Mathematical Olympiad.

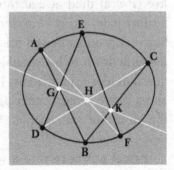

It is likely that Pascal conjured up his notions of infinity during his work on a new field of mathematics called projective geometry. That subject deals with the properties of shapes that are unchanged when they are projected onto other surfaces, like an object's shadow on the floor. One of the concepts of projective geometry is "the point at infinity," which arises in perspective drawing when one imagines, for example, extending a narrowing street indefinitely, until the parallel lines appear to meet. Although "the point at infinity" does not exist in the physical world (and certainly not in Pascal's knowledge of the physical world), it can be imagined.

After his father died in 1650, Pascal inherited a good deal of money, and he continued to mix with the best of society, as befitting his wealth. For a time, he kept a coach with six horses. Living in fashionable society and attending the various salons of Paris, Pascal was a man of the world. Paradoxically, he also became involved with an ascetic religious sect known as Jansenism, named after Cornelius Jansen, bishop of Ypres. The Jansenists were as strict as the Puritans, obsessed with original sin, human depravity, and predestination. T. S. Eliot once described Pascal as "a man of the world among ascetics, and an ascetic among men of the world; he had the knowledge of worldliness and the passion of asceticism, and in him the two are fused into an individual whole."

In addition to his scientific work, Pascal is known for his influential but unfinished *Pensées* (Thoughts), a collection of fragments on theology and philosophy, and often biting quips about other intellectuals of the day. Sickly for much of his life, Pascal died in August 1662, most probably of stomach cancer. In his later years, he commented that "sickness is the natural state of Christians...one hour's pain is a better teacher than all the philosophers put together."

What interests me most about Blaise Pascal is his imagination of the infinities—the infinitely small and the infinitely large—and the convolution of human beings with those unattainable realms. Certainly, Christian religious thinkers from St. Augustine on had discussed the infinite power of God. But there was no evidence of any physical thing remotely approaching the hugely small or the hugely large. Evidently, Pascal was mentally following the narrowing street, on and on in the world of his imagination. Scientists today have done the same. And with new discoveries in physics and astronomy—discoveries that not even Pascal could have imagined—we have found astonishing limits to the large and the small. Not mundane limits caused by the inadequacy of our measuring devices, but fundamental limits caused by the nature of time and space.

First, the large. A giant being, gazing at the cosmos, would see outer space as a vast dark sea, mostly empty but punctuated by glowing islands of light, the galaxies. On average, each galaxy, like our Milky Way, contains about a hundred billion stars and is about a thousand billion times the size of a star. Astronomers have, in fact, measured distances out to several hundred thousand galaxy diameters, the farthest known regions of reality. Space could extend far beyond even that great distance, but we can never see it for a fascinating reason. Since the 1920s, we have observed with our giant telescopes that the universe is expanding, with the galaxies moving away from one another like points painted on an expanding balloon.

If that picture is played backward, the matter of the universe hurtles together until a definite time in the past, about fourteen billion years ago, when everything was crammed together into a region of enormously high density and temperature, the so-called Big Bang beginning of our universe. The universe could be infinite in extent, but we cannot see beyond a certain distance because there hasn't been enough time since the Big Bang for light to have traveled from there to here. It is as if we found ourselves in a vast dark palace, with unlit chandeliers covering the ceiling, and suddenly the lights are turned on (the Big Bang). In the first few moments, we would see only the nearest chandeliers, because the light from the more distant ones has not yet reached our eyes. As time passed, we would see more and more distant parts of the palace. But at any given time, there would be an outer region beyond which we couldn't yet see. Thus, in our search for Pascal's infinity of the large, we reach a limit brought about by the finite age of our universe and the finite speed of light.

Now, the infinity of the small. Like the cosmos at large, atoms are mostly empty space. In each atom, a tiny nugget at the center, called the nucleus, is surrounded by electrons, almost weightless by comparison and orbiting at a distance a hundred thousand times the size of the nucleus. Let us travel to smaller sizes. The atomic nucleus is divisible, composed of smaller particles called protons and neutrons. And each of those is made of even smaller particles, called quarks, whose sizes were first measured with giant particle accelerators in 1969. A quark is about one hundred million times smaller than an atom.

Are quarks the end of the line, the smallest objects in nature? If he were alive today, Pascal would say no. He would simply imagine cutting a quark into two, then cutting each of those pieces in two, and so on, ad infinitum. If we follow through with this Pascalian prescription, however, we eventually hit another limit. We reach a point where gravitational physics and quantum physics come together in an unholy marriage. Gravitational physics, described by Einstein's theory of general relativity, tells us that the geometries of space and time are affected by mass and energy. That is, a mass like the Sun bends space the same way that a bowling ball on a trampoline sinks and flexes the mat beneath it. Masses also make time flow more slowly the closer you are to the mass.

The other partner in the marriage is quantum physics. Quantum physics, also developed in the 1920s, shows that in the subatomic realm, particles take on a hazy, nondefinite character, behaving as if they existed in several places at once. Although we don't yet have a theory of "quantum gravity," we can still estimate the size of the region in which quantum physics and gravitational physics would merge. This ultra-tiny scale is called the "Planck length," named after the physicist Max Planck, a pioneer in quantum physics. The Planck length is 10^{-33} centimeters, a hundred million billion times smaller than a quark. Another way to visualize the infinitesimal size we are talking about: the Planck length is smaller than the nucleus of an atom by about the same ratio as the nucleus is smaller than the state of Rhode Island. It staggers the mind that we have anything at all to say about such infinitesimal elements of reality.

Because of the hazy, nondefinite character of quantum

physics (called the Heisenberg uncertainty principle), at the dimensions of the Planck length, space and time churn and seethe, with the distance between any two points wildly fluctuating from moment to moment, and time randomly speeding and slowing, perhaps even going backward and forward. In such a situation, time and space no longer exist in a way that has meaning to us. The sensation of smooth time and space that we experience in our large world of houses and trees results only from averaging out this extreme lumpiness and chaos at the Planck length, in the same way that the graininess of a beach disappears when looked at from a thousand feet up.

Thus, if we relentlessly halve space again and again, paying homage to Pascal, searching for the infinity of the small, once we arrive at the phantasmagoric world of Planck, space no longer has meaning. Instead of probing the nature of the infinity of the small, we have invalidated the words used to ask the question. Space has been blown thin by an ancient glassblower, so thin that it dissolves into nothingness. The Planck world is a ghost world. It is a world without "time" and without "space." Just as Pascal suggested, we find ourselves at the abyss between nothingness and infinity. And in doing so, we have found limits to the smallest and largest things observable, limits imposed by the science of two and a half centuries after Pascal.

Scientists today, especially physicists, have also reached a point where their imagination far exceeds the possibility of experimental testing. Physicists have hypothesized that the smallest elements of nature are not particles, like electrons, but extremely tiny one-dimensional "strings" of energy, the size of the Planck length—a size that would require particle accelerators larger than the Earth to explore. Physicists have also speculated about the existence of other universes, possibly infinite in extent, which will never be in contact with our universe and thus are impossible to confirm. Cosmologists have theorized about the origin of our universe. Did time and space begin with the Big Bang, or did they exist before, in some quantum haze? Although there are various theories to answer these questions, it is unlikely that we will ever know which, if any, is correct. In sum, we have added much detail to Pascal's two infinities, we have embroidered his imagination with more advanced imagination, but we still find ourselves in the realm of the hypothetical as opposed to the definite, and we may remain there for a long time, perhaps an infinite time. A great philosopher of science, Karl Popper, once said that a proposition is not scientific unless it can be falsified—that is, unless one can perform an experiment that proves it wrong. At any given moment in history, the scientific theories and ideas we endorse are those that have not been falsified. If we can never test the infinities of the small and the infinities of the large, perhaps these notions are not scientific after all. But they are certainly vibrant in the realm of the imagination.

Finally, I would like to return to the invocation of human beings in Pascal's passage from the *Pensées*—the physical aspects, the philosophical, and the psychological. Pascal could simply have said that the universe extends to infinity, both in the large and the small. But he refers to cosmic scales in terms of human beings. First, compared to the infinitely large: "What is man in the infinite?" And then, compared to the infinitely small: "For who will not be astounded at the fact that our body, which a little while ago was imperceptible in the universe... is now a colossus, a world, or rather a whole in respect of the nothingness [smallness] which we cannot reach?" Man has a "body given him by nature between these two abysses of the Infinite and Nothing."

Using our modern knowledge of the sizes of things, we can say very specifically where human beings fit in the hierarchy of the cosmos. How many times should the size of a human body be *halved* to reach the size of an atom (a size unknown until the twentieth century)? The answer is about 33. Going in the opposite direction, one can ask how many times the size of the human body should be *doubled* to reach the size of a typical star, like our Sun, the largest object known by Pascal. The answer is 30. Thus, counting in doublings, the size of a human being is nearly halfway between an atom and a star—certainly not two infinities, but a minuscule thing of the natural world at one end and a gargantuan at the other. So, although Pascal did not have the quantitative knowledge of the cosmos we have today, there is a sense in which we human beings, at least physically, are indeed between the large and the small.

More interesting, perhaps, are the psychological and even theological tones in the passage: "He who regards himself in this light will be afraid of himself... and will tremble at the sight of these marvels... [Man] is equally incapable of seeing the Nothing from which he was made, and the Infinite in which he is swallowed up." As mentioned earlier, Pascal was extremely devout, even in the context of his time and place. Undoubtedly, in these sentences Pascal was referring to the insignificance and limitations of Man in the divine sensorium of God. The "Nothing" here probably refers to divine Creation—both the creation of human beings and the creation of the universe as a whole. Man's incapability of fathoming nothingness and infinity, known only to God, reminds me of the passage in Milton's *Paradise Lost*, published a mere five years after Pascal's death, in which Adam questions the angel Raphael about celestial mechanics. Raphael offers some vague hints and then says that "the rest / From Man or Angel the great Architect / Did wisely to conceal, and not divulge / His secrets to be scann'd by them who ought / Rather admire."

It is clear that there are boundaries to humankind's knowledge. But I disagree with Pascal that we human beings should fear what we do not understand, the infinities on both sides of us. Certainly there are fundamental limits to exploring the large and the small, as discussed above. But are we to "tremble" at their contemplation? Are we to bemoan our inability to grasp such things? Einstein once wrote, "The most beautiful experience we can have is the mysterious. It is the fundamental emotion which stands at the cradle of true art and true science." By the "mysterious," I do not think Einstein was referring to something fearful or supernatural. I believe he was speaking about the boundary between the known and the unknown. Standing at that boundary is an exhilarating experience. And it is a deeply human experience—concerning what the human mind understands and what that mind does not *yet* understand. The boundary between the known and the unknown is not a static boundary. It moves as we acquire new knowledge and understanding. Five hundred years ago, we did not understand the nature of heat or electricity. A hundred years ago, we did not understand the mechanism by which living organisms provide instructions to create descendant living organisms. The boundary between the known and the unknown constantly shifts. The other side is the "mysterious." That other side intrigues us, it stimulates us, it provokes us, it haunts us. And it produces new science, and new art.

NOTHINGNESS

What Came Before the Big Bang?

On Wednesday, February 11, 1931, Albert Einstein met for more than an hour with a small group of American scientists in the cozy library of the Mount Wilson Observatory, near Pasadena, California. The subject was cosmology, and Einstein was poised to make one of the more momentous statements in the history of science. With his theories of relativity and gravity long confirmed and his Nobel ten years old, he was at this time the most famous scientist in the world. ("Photographers lunged at me like hungry wolves," he wrote in his diary when his ship landed in New York two months earlier.)

For years, Einstein had insisted, like Aristotle and Newton before him, that the universe was a magnificent and immortal cathedral, fixed for all of eternity. In this picture, time runs from the infinite past to the infinite future, but little changes over time. Einstein dismissed the evolving cosmology of a Russian physicist as formally correct but of no physical significance. When a prominent Belgian scientist proposed in 1927 that the universe was growing in size like an expanding balloon, Einstein pronounced the idea "abominable."

Recently, however, the great physicist had been confronted with telescopic evidence that the distant galaxies were in flight. Perhaps even more convincing to him, his mathematical model for a static universe had been shown to be like a pencil balanced on its point: give it a tiny nudge and it starts to move. By the time he arrived in Pasadena, Einstein was ready to acknowledge a cosmos in flux. In his thick German accent, he told the surrounding men in their suits and ties that the observed motion of the galaxies "has smashed my old construction like a hammer blow." Then he swung down his hand to emphasize the point. What rose in the shards of that hammer blow was the Big Bang cosmology: the universe is not static and everlasting; rather, it "began" some fourteen billion years ago and has been expanding ever since. According to current data, our universe will keep expanding forever.

Sean Carroll, a professor of physics at the California Institute of Technology, is a Big Bang cosmologist. But more than that, he is one of a small platoon of physicists who call themselves "quantum cosmologists." He wants to know what happened at the very beginning, and maybe even *before* that. Carroll and other quantum cosmologists believe that not only the universe was created at the Big Bang, but perhaps time itself. With pencil and paper, these theoretical physicists are investigating what, if anything, existed before the Big Bang, whether time had a beginning, and why we can tell the future from the past. Such bedrock questions in physics, seriously posed only recently, might be likened to Descartes asking for proof of his existence. Such questions are also related to Pascal's notion that we, and the universe, emerged from "nothingness." According to modern cosmologists, the entire observable universe was once microscopic in size. Thus Pascal's idea of the infinitely small, his "nothingness," might be associated with the origin of our universe.

Quantum cosmology is speculative work. For one thing, the birth of our universe was a one-performance event, and we weren't there in the audience. But more importantly, an understanding of the very beginning requires a knowledge of gravity at enormously high densities of matter and energy, so-called quantum gravity, discussed in the last chapter. Physicists believe that in this quantum era, the *entire universe* we see today was far smaller than a single atom—roughly a million billion billion times smaller (assuming the universe went through an inflationary epoch). The temperature was nearly a million billion billion billion degrees. And time and space churned like boiling water. Of course, such things are unimaginable. But theoretical physicists try to imagine them with pencil and paper and mathematics. Somehow, time as we know it emerged in that fantastically dense nugget. Or perhaps time already existed, but what appeared was the arrow of time, the direction toward the future.

Physicists hope that within the next fifty years or so, string theory or other new theoretical work will provide a good understanding of quantum gravity, including an explanation of how the universe began. Until then, some of the deepest minds in physics, including Stephen Hawking and Andrei Linde and Alexander Vilenkin, have debated different hypotheses, each backed up with pages of calculations. But it is a tiny field, not for the timid. Carroll explained to me its allure: "high risk, high gain." And down the rabbit hole we go.

When I reached Sean Carroll by Skype, he was wearing a hoodie and jeans in the comfortable study of his home in Los Angeles. I was stationed in an uninhabitable guest room of my home in Concord, Massachusetts—practically next door on the scale of a galaxy. Carroll was relaxed as he talked about his favorite subject. Forty-nine years old, barrel-chested with a full head of reddish hair, puffy cheeks and jowls, a mischievous schoolboy glint in his eye, Carroll is an articulate explicator of science as well as a well-regarded physicist. He's written scientific papers with titles like "What If Time Really Exists?" and popular books such as *From Eternity to Here: The Quest for the Ultimate Theory of Time*. He quotes from people like Parmenides and Heraclitus.

Carroll is obsessed with the relative smoothness and order of the universe. Order in physics has a precise meaning. It can be quantified. Furthermore, conditions of disorder are more probable than conditions of order, just as a deck of cards, once shuffled, is more likely to be found with the cards jumbled up than with the cards arranged by number and suit. Applying those considerations to the cosmos at large, physicists expect that given the amount of matter in the observable universe, we would expect it to be far more disordered and lumpy than it actually is. To be more exact, our observable universe has something like a hundred billion galaxies in it, which, when viewed over sufficiently large expanses of space, look as smooth as a pebbly beach seen from afar. Any large volume of space looks about like any other large volume. But it would be far more probable, say the physicists, to see that same material concentrated in a much smaller number of ultra-large galaxies or large clusters of galaxies or perhaps even in a single massive black hole—analogous to all the sand on a beach concentrated in a few silicon boulders.

The improbable smoothness of the observable universe, in turn, points toward unusually tidy conditions near the Big Bang. We don't understand why. But it's a clue. Not a shrinking violet with his cosmological opinions, Carroll told me, "I strongly believe that the low entropy [i.e., high order and smoothness] of the early universe is a puzzle that the wider cosmology community doesn't take nearly as seriously as they should. Misunderstandings like that offer opportunities for making new breakthroughs." Carroll and other physicists believe that order is intimately connected to the "arrow" of time. In particular, the forward direction of time is determined by the movement of order to disorder. For example, a movie of a glass goblet falling off a table and shattering on the floor would look normal to us; if we saw a movie of scattered shards of glass jumping off the floor and gathering themselves into a goblet perched on the edge of the table, we would say that movie was being played backward in time. Likewise, clean rooms left unattended become dusty with time, not cleaner. What we call the "future" is the condition of increasing mess; what we call the "past" is increasing tidiness. Our ability to easily distinguish between the two shows that our world has a clear direction of time. (Only theoretical physicists worry about such things.) So too in the cosmos at large. Stars radiate heat and light, slowly extinguish their nuclear fuel, and finally turn into cold cinders drifting through space. Never does the reverse happen.

Which brings us back to the unusual orderliness of our universe. Working with Alan Guth, a pioneering cosmologist at the Massachusetts Institute of Technology, Carroll has developed a not-yet-published theory called "Two-Headed Time." In this theory, time has existed forever. But unlike in the static models of Aristotle and Newton and Einstein, the universe changes as the eons go by. Furthermore, the evolution of the cosmos is symmetric in time, with the behavior of the universe before the Big Bang a nearly mirror image of its behavior after the Big Bang. Until fourteen billion years ago, the universe was contracting. It reached a minimum size at the Big Bang (which we call $t = 0$) and has been expanding ever since, like a Slinky that falls to the floor, reaches a maximum compression upon impact, and then bounces back to larger dimensions. A few other quantum cosmologists have proposed related models. Because of unavoidable random fluctuations required by quantum physics, the contracting universe would not be an exact mirror image of the expanding universe, so that a physicist named Alan Guth probably did not exist in the contracting phase of our universe. But the before and after would look extremely similar.

Now it is well known in the science of order and disorder that, other things being equal, larger spaces allow for more disorder, essentially because there are more places to scatter things. Equivalently, smaller spaces have more order. As a consequence, in the Carroll-Guth picture, the order of the universe was at a *maximum* at the Big Bang, with order decreasing both before and after. Recall that the forward direction of time is the movement of order to disorder. Thus the future points away from the Big Bang in both directions of time. A person living in the contracting phase of the universe sees the Big Bang in her past, just as we do. When she dies, the universe is larger than when she was born, just as it will be for us.

If you think of time as a long road and the Big Bang as a pothole somewhere in that road, then a sign at the pothole telling you the direction to the future would have two arrows pointing in opposite directions. Hence the name "Two-Headed Time." Near the pothole itself, caught between one arrow pointing one way and another arrow pointing the opposite way, time would have no clear direction. Time would be confused. In the subatomic version of goblets and houses, shards of glass would jump off the floor to form goblets as often as those goblets would fall from tables and shatter. Unattended houses would become neater as often as they would become more cluttered. Both movies would be equally familiar to any subatomic being living at that moment.

Science fiction? Perhaps. Perhaps not. Right or wrong, these ideas are profound. Says Carroll: "When I came to understand that the reason I can remember the past but not the future is ultimately related to conditions at the Big Bang, that was a startling epiphany."

Another major proposal is that the universe, and time, did not exist before the Big Bang. Time emerged. Advocates of this hypothesis believe that the universe materialized literally out of nothing, at a tiny but finite size, the Planck size, and expanded thereafter. Such things are possible in quantum physics. But time didn't exist at that time. There were no moments prior to the moment of smallest size because there was no "prior." Likewise, there was no "creation" of the universe, since that concept implies action in time. As Hawking describes it, "the universe was neither created nor destroyed. It would just BE." Such notions as existence and being in the absence of time are not fathomable within our limited human experience. We don't even have language to describe them. Nearly every sentence we utter has some notion of *before* and *after*.

One of the first quantum cosmologists to suggest that the universe could appear out of nothing was Alexander Vilenkin, a Ukrainian scientist who came to the United States in 1976, in his midtwenties, to do graduate work and is now professor of physics at Tufts University, near Boston. When I visited him in his new office on a hot day in July, he was wearing sandals and a loose black shirt. The single window looked out on a dull brick building across the street. "The view from my previous office was better," he told me. Boxes of unpacked books littered the floor; on his bookshelf was an Einstein doll given him by his daughter.

Before coming to the United States, Vilenkin saw his acceptance to graduate school in the Soviet Union rescinded, possibly because of the KGB. So he began work as a night watchman in a zoo—giving him plenty of time to think cosmological thoughts. In the United States, Vilenkin got his PhD in biophysics, not cosmology. "I was doing cosmology on the side," he says. "It was not a reputable field of research at that time." Vilenkin is a serious man who, unlike many physicists, does not joke around much, and he takes his work on the universe at t = 0 extremely seriously. "No cause is required to create a universe from quantum tunneling," he says, "but the laws of physics should be there." Briefly, we chatted about what "there" means when time and space do not yet exist. And how did the laws of physics get "there"? On this score, Vilenkin likes to quote St. Augustine, who was often asked what God was doing before He created the universe. In his *Confessions*, Augustine replies that since God created time when He created the universe, there was no "before," and no "then." Blaise Pascal, a devout Catholic, would have shared Augustine's view. His "nothingness" referred not only to the infinitely small, but also to conditions at God's Creation event.

When Vilenkin talks about "quantum tunneling," he is referring to the spooky phenomenon in quantum physics in which objects can perform such magic feats as passing through a mountain and suddenly appearing on the other side, without ever going over the top. That mystifying ability, which has been verified in the laboratory, follows from the fact that subatomic particles behave as if they could be in many places at once. Quantum tunneling is common in the tiny world of the atom but totally negligible in our human world—explaining why the phenomenon seems so absurd. But in the quantum era of cosmology, very near t = 0, the *entire universe* was the size of a subatomic particle. Thus, the entire universe could have "suddenly" appeared from wherever things originate in the impossible-to-fathom quantum haze of probabilities. (I put "suddenly" in quotation marks because time didn't exist; but I have just now realized that in this very sentence I used the verb "did," which is the past tense of "do," and "now," which...)

What does it mean to say that the entire universe was like a subatomic particle, existing in the twilight world of the quantum? James Hartle, a leading quantum cosmologist at the University of California at Santa Barbara, has developed with Hawking one of the most detailed models of the universe "during" the quantum era near the Big Bang. Time appears nowhere in Hartle and Hawking's equations. Instead, they compute the probability of certain snapshots of the universe, using quantum physics to do so.

Although a world's expert in quantum theory, Hartle admits to being baffled in the application of quantum physics to the universe as a whole. "It is a mystery to me," he told me, "why we have quantum mechanics when there is only one state of the universe." In other words, why should there be probabilities of alternative conditions of our universes when we inhabit only one condition? And do those other potential conditions actually exist in other universes somewhere?

The quantum cosmologists are not unaware of the vast philosophical and theological reverberations of their work. As Hawking says in his book *A Brief History of Time*, many people believe that God, while permitting the universe to evolve according to fixed laws of nature, was uniquely responsible for winding up the clock at the beginning and choosing how to set it in motion. Hawking's own theory provides an explanation for how the universe might have wound itself up—by proposing a method of calculating the "early" snapshots of the universe that has no dependence on "initial conditions" or boundaries or anything outside the universe itself.

The icy rules of quantum physics are completely sufficient. "What place, then, for a Creator?" asks Hawking. Reaching a similar conclusion, physicist Lawrence Krauss wrote an entire book, titled *A Universe from Nothing*, in which he argues that advances in quantum cosmology show that God is irrelevant at best.

Of course, one would expect most quantum cosmologists to be atheists, like the majority of scientists. A prominent exception is Don Page, a leading quantum cosmologist at the University of Alberta, and also an evangelical Christian. Page is a master computationalist. When he and I were fellow graduate students in physics at the California Institute of Technology, he would quietly take out a fine-point pen whenever confronted with a difficult physics problem and, without flinching, begin scribbling one equation after another in a dense jungle of mathematics until he had arrived at the answer. Although he collaborated with Hawking on major papers, Page parts ways with him on the subject of God. As Page recently told me, "As a Christian, I think there is a Being outside the universe that created the universe and caused all things. God is the true Creator. All of the universe is caused by God." And in a guest column in Sean Carroll's blog (which is called *The Preposterous Universe*), Page sounded like a scientist and a theist simultaneously: "One might think that adding the hypothesis that the world (all that exists) includes God would make the theory for the entire world more complex, but it is not obvious that is the case, since it might be that God is even simpler than the universe, so that one would get a simpler explanation starting with God than starting with just the universe."

Significantly, most quantum cosmologists do not believe

that anything *caused* the creation of the universe. As Vilenkin said to me, quantum physics can produce a universe without cause—just as quantum physics shows how electrons can change orbits in atoms without cause. There are no definite cause-and-effect relationships in the quantum world, only probabilities. Carroll put it this way: "In everyday life we talk about cause and effect. But there is no reason to apply that thinking to the universe as a whole. I do not feel in any way unsatisfied by just saying 'that's the way it is.'"

The notion of an event or state of being without cause drives hard against the long grain of science. For centuries, science has attempted to explain all events as the logical consequence of prior events. Page argues that at the origin of our universe—whether in the Two-Headed Time model, where time pauses without direction, or in the universe-out-of-nothing model—there was no clear distinction between cause and effect. If causality can dissolve in the quantum haze of the origin of the universe, Page and other physicists question its solidity even in the world that we live in, long after the Big Bang—surely part of the same reality. "Causality within the universe is not fundamental," says Page. "It is an approximate concept derived from our experience with the world." Strict causality could be an illusion, a way for our brains, and our science, to make sense of the world.

Now, we are struck. A crack in the marble foundation of causality sends tremors into philosophy, religion, ethics, and more. For example, without strict causality, how do we humans make decisions? What are the relative roles of prior events and conditions versus sudden impulse or even simply action without cause? What happens to accountability? Decision making is such a delicate and complex mental process. If causality is only approximate, we don't know where the tipping point lies, where the decision is so fragile that it appears without definite cause.

Quantum cosmology has led us to question the most fundamental aspects of existence and being, questions that we rarely ask. Most of us aim in our short century or less to create a comfortable existence within the tiny rooms of our lives. We eat, we sleep, we get jobs, we pay the bills, we have lovers and children. Some of us build cities or make art. But with the luxury of true freedom of mind, there are larger concerns. Look at the sky. Does space go on forever, to *infinity*? Or is it finite but without boundary or edge, like the surface of a sphere? Either answer is disturbing, and unfathomable. Where did our Sun and Earth come from? Where did we come from? Quickly, we realize how limited we are in our experience of the world. What we see and feel with our bodies, caught midway between atoms and stars, is but a small swath of the spectrum, a sliver of reality.

In the 1940s, the American psychologist Abraham Maslow developed the concept of a hierarchy of human needs, starting with the most primitive and urgent, and ending with the most lofty and advanced for those fortunates who had satisfied the baser needs. At the bottom of the pyramid are physical needs for survival, like food and water. Next up is safety. Higher up is love and belonging, then self-esteem, and finally self-actualization. This highest of Maslow's proposed needs is the desire to get the most out of ourselves, to be the best we can be. I would suggest adding one more category at the very top of the pyramid, even above self-actualization: imagination and exploration. The need to imagine new possibilities, the need to reach out beyond ourselves and understand the world around us. Wasn't that need part of what propelled Marco Polo and Vasco da Gama and Einstein? Not only to help ourselves with physical survival or personal relationships or self-discovery, but to know and comprehend this strange cosmos we find ourselves in. The need to explore the really big questions asked by the quantum cosmologists. How did it all begin? Far beyond our own lives, far beyond our community or our nation or planet Earth or even our solar system. How did the universe begin? It is a luxury to be able to ask such questions. It is also a human necessity.

On Nothingness

Nothing will come of nothing.

—WILLIAM SHAKESPEARE, *KING LEAR* (1606)

Man is equally incapable of seeing the nothingness from which he emerges and the infinity in which he is engulfed.

—BLAISE PASCAL, "THE MISERY OF MAN WITHOUT GOD," *PENSÉES* (1670)

The . . . "lumniferous ether" will prove to be superfluous as the view to be developed here will eliminate [the condition of] absolute rest in space.

—ALBERT EINSTEIN, *ON THE ELECTRODYNAMICS OF MOVING BODIES* (1905)

As we have struggled through the ages to fathom this strange and wondrous cosmos in which we find ourselves, few ideas have been richer than the concept of nothingness. For to understand anything, as Aristotle argued, we must understand what it is not. To understand matter, said the ancient Greeks, we must understand the "void," or the absence of matter. Indeed, in the fifth century BC, Leucippus argued that without the void there could be no motion because there would be no empty spaces for matter to move into. According to Buddhism, to understand our ego we must understand the ego-free state of "emptiness," called *śūnyatā*. To understand the civilizing effects of society, we must understand the behavior of human beings removed from society, as William Golding so powerfully explored in his novel *Lord of the Flies*.

Following Aristotle, let me say what nothingness is not. It is not a unique and absolute condition. Nothingness means different things in different contexts. From the perspective of life, nothingness might mean death. To a physicist, it might mean the complete absence of matter and energy (an impossibility, as we will see), or even the absence of time and space. To a lover, nothingness might mean the absence of his or her beloved. To a parent, it might mean the absence of children. To a theologian or philosopher like Pascal, nothingness meant the infinitely small, as well as a timeless and spaceless domain known only by God. When King Lear says to his daughter Cordelia, "Nothing will come of nothing," he means that she will receive far less of his kingdom than her two fawning sisters unless she can express her boundless love for him. The second "nothing" refers to Cordelia's silence contrasted with her sisters' gushing adoration, while the first is her impending one-room shack compared to their opulent palaces. These negations, of course, leave a lot of room for the meaning of nothingness, and there are many such meanings.

My own most vivid encounter with nothingness occurred not while dividing up my kingdom or while contemplating the absence of three-dimensional space in quantum physics, but in a remarkable experience I had as a nine-year-old child. It was a Sunday afternoon. I was standing alone in a bedroom of my home in Memphis, Tennessee, gazing out the window at the empty street, listening to the faint sound of a train passing a great distance away, and suddenly I felt that I was looking at myself from outside my body. For a brief few moments, I had the sensation of seeing my entire life, and indeed the life of the entire planet, as a brief flicker in a vast chasm of time, with an infinite span of time before my existence and an infinite span of time afterward. My fleeting sensation included infinite space. Without body or mind, I was somehow floating in the gargantuan stretch of space, far beyond the solar system and even the galaxy, space that stretched on and on and on. I felt myself to be a tiny speck, insignificant. A speck in a huge universe that cared nothing about me or any living beings and their little dots of existence. A universe that simply was. And I felt that everything I had experienced in my young life, the joy and the sadness, and everything that I would later experience, meant absolutely nothing in the grand scheme of things. It was a realization both liberating and terrifying at once. Then the moment was over, and I was back in my body.

The strange hallucination lasted only a minute or so. I have never experienced it since. Although nothingness would seem to exclude awareness along with everything else, awareness was part of that childhood experience, but not the usual awareness I would locate within the three pounds of gray matter in my head. It was a different kind of awareness. I am not religious, and I do not believe in the supernatural. I do not think for a minute that my mind actually left my body. But for a few moments I did experience a profound absence of the familiar surroundings and thoughts we create to anchor our lives. It was a kind of nothingness. Perhaps not Pascal's nothingness, but a personally experienced nothingness.

Although nothingness may have different meanings in different circumstances, I want to emphasize what is perhaps obvious: All of its meanings involve a comparison to a material thing or condition we know. That is, nothingness is a *relative* concept. We cannot conceive of anything that has no relation to the material things, thoughts, and conditions of our existence. For example, sadness, by itself, has no meaning without reference to joy. Poverty is defined in terms of a minimum income and standard of living. The sensation of a full stomach exists in comparison to that of an empty one. In nature it is the *difference* in adjacent conditions that makes things happen. An airplane is kept aloft by the difference in air pressure below and above its wings. Make the pressures the same, whatever the value, and the plane cannot fly. Steam engines are driven by the difference in temperature between the boiler and surrounding material. Make the temperature the same everywhere, whatever the value, and the engine will come to a halt. Is a person tall, or heavy, or intelligent? Tall compared to what? Intelligent compared to what? Absolute values are meaningless. Similarly, nothingness has meaning only when compared to something.

My first experience with nothingness in the material world of science occurred when I was a graduate student in theoretical physics at the California Institute of Technology.

In my second year, I took a formidable course with the title of Quantum Field Theory, which explained how all of space is filled up with "energy fields." There is an energy field for gravity and an energy field for electricity and magnetism, and so on. What we regard as physical "matter" is the excitation of the underlying energy fields. A key point is that according to the laws of quantum physics, all of these energy fields are constantly jittering a bit—it is an impossibility for an energy field to be completely dormant—and the jittering causes sub-atomic particles like electrons and photons to appear for a brief moment and then disappear again, even when there is no persistent matter. Physicists call a region of space with the lowest possible amount of energy in it the "vacuum." But the vacuum cannot be free of energy fields. The energy fields necessarily permeate all space. And because they are constantly jittering, they are constantly producing matter, at least for brief periods of time. Thus the "vacuum" in modern physics is not the void of the ancient Greeks. The void does not exist. (Pascal's "vacuum" might have been close to the physicists' vacuum.) Every cubic centimeter of space in the universe, no matter how empty it seems, is actually a chaotic circus of fluctuating energy fields and particles flickering in and out of existence on the subatomic scale. Thus, at the material level, there is no such thing as nothingness.

Remarkably, the active nature of the "vacuum" has been observed in the lab. In the 1920s, physicists discovered that electrons (the smallest subatomic particles that carry electrical charge) are constantly spinning like little tops. Unlike ordinary tops, every electron has an identical amount of spin. Circulating electrical charge creates magnetism, so all electrons are identical tiny magnets in addition to being tiny tops. And just as the spin axis of a spinning top, when tilted relative to the direction of gravity, will precess (slowly rotate) about the vertical, an electron will precess when tilted relative to the direction of a magnetic field. The rate of precession can be measured to high accuracy, and that rate, in turn, is determined by the magnetic strength of electrons. Now comes the role of the quantum vacuum. If space were completely empty, the magnetic strength of the electron would be predicted to be exactly 1, in the appropriate units. But the theory of quantum physics proclaims that the electric fields in the vacuum are constantly producing massless particles called photons, which interact with all electrically charged particles and alter their properties. These ghostlike photons pop out of the vacuum into being, enjoy their lives for perhaps a billionth of a billionth of a second, and then disappear again. During their brief moment of existence, they collide with electrons and slightly change their magnetic strength. That strength has been measured in the laboratory to be 1.00115965221. By comparison, complex and highly mathematical equations from the theory of the quantum vacuum (spelled out in my grad school textbook on quantum field theory) *predict* that the magnetic strength of the electron should be 1.00115965246—a fantastic validation of the quantum theory of the vacuum. It is a triumph of the human mind to understand so much about empty space.

The concept of empty space—and nothingness—played a major role in modern physics even before our understanding of the quantum vacuum. According to findings in the mid-nineteenth century, light is a traveling wave of electromagnetic energy, and it was conventional wisdom that all waves, such as sound waves and water waves, required a material medium to carry them along. Take the air out of a room, and you will not hear someone speaking. Take the water out of a lake, and you cannot make waves. The material medium hypothesized to convey light was a gossamer substance called the "ether." Because we can see light from distant stars, the ether had to fill up all space. Thus, there was no such thing as empty space. Space was filled with the ether.

In 1887, in one of the most famous experiments in all of physics, two American physicists at what is now Case Western Reserve University in Cleveland, Ohio, attempted to measure the motion of the Earth through the ether. Their experiment failed. Or, rather, they could not detect any effects of the ether. In 1905, a twenty-six-year-old patent clerk named Albert Einstein proposed that the ether did not exist. Instead, he hypothesized that light, unlike all other known waves, could propagate through completely empty space. All this was before quantum physics.

That denial of the ether, and hence embrace of a true emptiness, followed from a deeper hypothesis of the young Einstein: There is no condition of absolute rest in the cosmos. Without absolute rest, there cannot be absolute motion. You cannot say that a train is moving at a speed of 50 miles per hour in any absolute sense. You can say only that the train is moving at 50 miles per hour relative to another object, like a train station. Only the *relative* motion between two objects has any meaning. The reason Einstein did away with the ether is that it would have established a reference frame of absolute rest in the cosmos. With a material ether filling up all space, you could say whether an object is at rest or not (relative to the ether), just as you can say whether a boat in a lake is at rest or in motion with respect to the water. So, through the work of Einstein, the idea of material emptiness, or nothingness, was connected to the rejection of absolute rest in the cosmos. In sum, first there was the ether filling up all space. Then Einstein removed the ether, leaving truly empty space. Then other physicists filled space again with quantum energy fields. But the quantum energy fields do not restore a reference frame of absolute rest because they are not static materials in space. Einstein's principle of relativity remained.

One of the pioneers of quantum field theory was the legendary physicist Richard Feynman, a professor at Caltech and a member of my PhD thesis committee. In the late 1940s, Feynman and others developed the theory of how electrons interact with the ghostly photons of the vacuum. Earlier in that decade, as a cocky young scientist, he had worked on the Manhattan Project. By the time I knew him at Caltech, in the early 1970s, Feynman had mellowed a bit but was still ready to overturn received wisdom at the drop of a hat. Every day, he wore white shirts, exclusively white shirts, because he said they were easier to match with different colored pants, and he hated to spend time fussing about his clothes. Feynman also had a strong distaste for philosophy. Although he had quite a wit, he viewed the material world in a highly straightforward manner, without caring to speculate on the purely hypothetical or subjective. He could and did talk for hours about the behavior of the quantum vacuum, but he would not waste a minute on philosophical or theological considerations of nothingness. My experience with Feynman taught me that a person can be a great scientist without concerning him- or herself with questions of "Why?"—which fall beyond the scientifically provable. However, Feynman did understand that the mind can create its own reality. That understanding was revealed when he gave the commencement address at my graduation from Caltech in 1974. It was a boiling day in late May, outdoors of course, and we graduates were all sweating heavily in our caps and gowns. In his talk, Feynman made the point that before publishing any scientific results, we should think of all the possible ways that we could be wrong. "The first principle," he said, "is that you must not fool yourself— and you are the easiest person to fool."

In the Wachowski Brothers' landmark film *The Matrix* (1999), we are well into the drama before we realize that all the reality experienced by the characters—the pedestrians walking the streets, the buildings and restaurants and nightclubs, the entire cityscape—is an illusion, a fake movie played in the brains of human beings by a master computer. Actual reality is a devastated and desolate planet, in which human beings are imprisoned, comatose, in leaf-like pods and drained of their life energy to power the machines. I would argue that much of what we call reality in our lives is also an illusion, or at least a view of reality largely shaped by our perceptions.

First, there is our own consciousness. As I will argue later in the chapter "Immortality," the magnificent and powerful experience of consciousness is not some transcendent quality, rising above the material world, but simply a *sensation* caused by the trillions of electrical and chemical flows within and between our neurons.

Likewise, our human-made institutions. We endow our art and our cultures and our codes of ethics and our laws with a grand and everlasting existence. We give these institutions an authority that extends far beyond ourselves. But in fact, I would suggest, all of these are constructions of our minds. That is, these institutions and codes and their imputed meanings are all mental constructions. They have no reality other than that which we give them, individually and collectively.

The Buddhists have understood this notion for centuries. It is part of the Buddhist concepts of emptiness and impermanence. The transcendent, nonmaterial, long-lasting qualities that we impart to other human beings and to human institutions are an illusion, like the computer-generated world in *The Matrix*. It is certainly true that we human beings have achieved what, to our minds, is extraordinary accomplishment. We have scientific theories that can make accurate predictions about the world. We have created paintings and music and literature that we consider beautiful and meaningful. We have entire systems of laws and social codes. But these things have no intrinsic value outside of our minds. And our minds are only collections of atoms, fated to disassemble and dissolve. For each of us, that will be the end of all consciousness and thought. And in that sense, we and our institutions are always approaching nothingness.

So where do such sobering thoughts leave us? Given our temporary and self-constructed reality, how should we then live our lives, as individuals and as a society? As I have been approaching my own personal nothingness, I have mulled these questions over quite a bit, and I have come to some tentative conclusions to guide my own life. Each person must think through these profound questions for him- or herself. There are no right answers. I believe that as a society we need to realize we have great power to make our laws and other institutions whatever we wish to make them. There is no external authority. There are no external limitations. The only limitation is our own imagination. So, we should take the time to think expansively about who we are and what we want to be.

As for each of us as individuals, until the day when we can upload our minds to computers, we are confined to our physical body and brain. And, for better or for worse, we are stuck with our personal mental state, which includes our personal pleasures and pains. Whatever concept we have of reality, without a doubt we experience personal pleasure and pain. We feel. Descartes famously said "I think, therefore I am." We might also say "I feel, therefore I am." And when I talk about feeling pleasure and pain, I do not mean merely physical pleasure and pain. Like the ancient Epicureans, I mean all forms of pleasure and pain: intellectual, artistic, moral, philosophical, and so on. All of these forms of pleasure and pain we experience, and we cannot avoid experiencing them. They are the reality of our bodies and minds, our internal reality. And here is the point I have reached: I might as well live in such a way as to maximize my pleasure and minimize my pain. Accordingly, I try to eat delicious food, to support my family, to create beautiful things, and to help those less fortunate than myself—because those activities bring me pleasure. Likewise, I try to avoid leading a dull life, avoid personal anarchy, and avoid hurting others because those activities bring me pain. That is how I should live. A number of thinkers far deeper than I, most notably the British philosopher Jeremy Bentham, have come to these same conclusions via very different routes.

What I feel and I know is that I am here now, at this moment in the grand sweep of time. I am not part of the void. I am not a fluctuation in the quantum vacuum. Even though I understand that someday my atoms will be scattered in soil and in air, that I will no longer exist, I am alive now. I am feeling this moment. I can see my hand on my writing desk. I can feel the warmth of the Sun through the window. And looking out, I can see a pine-needled path that goes down to the sea.

Atoms

It may have been the ancient Greeks who first conceived of a tiniest unit of matter, the atom, or *atomos*, meaning "uncuttable." Atoms were not only uncuttable. They were indestructible. Atoms protected us from the whimsy of the gods, said Democritus and Lucretius, because atoms could not be created or destroyed. Even the gods had to obey atoms. Newton also prized atoms, but as the handiwork of God rather than as a defense against Him. Newton, who understood the logic of nature better than any mortal before him, wrote: "It seems probable to me that God in the beginning formed matter in solid, massy, hard, impenetrable, moveable particles . . . so hard as never to wear or break in pieces; no ordinary power being able to divide what God Himself made *one* in the first creation." Indeed atoms were the ultimate oneness of the material world. Perfect in their indivisibility, perfect in their wholeness and indestructibility.

Atoms also unified the world, because a leaf and a human

being are made of the same atoms. Take apart a leaf or a human and we find identical atoms of hydrogen and oxygen and carbon and other elements. With atoms, we have a foundation for material reality. On that foundation, we can build *systems*. We can organize and construct the rest of the world. Said Lucretius: *Pleasing substances are made of smooth and round atoms, bitter substances of hooked and thorny atoms.* With atoms, we can make rules for the particular proportions in which different substances combine, as British chemist John Dalton did in the early nineteenth century. Carbon monoxide: one atom of carbon joined to one atom of oxygen. Carbon dioxide: one atom of carbon joined to two atoms of oxygen. Never carbon with one and a half atoms of oxygen. Because atoms cannot be divided. With atoms, we can predict the properties of the chemical elements, as Dmitri Mendeleev did in the mid-nineteenth century.

Atoms prevent us from falling forever into smaller and smaller rooms of reality, contrary to the notions of Pascal. When we reach atoms—so the thinking went—the falling stops. We are caught. We are safe. And from there, we begin our journey back up, building the rest of the world.

Although atoms had been conjectured for a couple of thousand years, their size wasn't known until the work of Albert Einstein, in his miracle year of 1905. Among many other things Einstein studied at the time—relativity, the particle nature of light, and so on—was the jittery motion of tiny particles suspended in a fluid, so-called Brownian motion, named after the botanist Robert Brown, who in 1827 first described the random dance of pollen suspended in water. Einstein reasoned that the jittery motion must be caused by collisions with water molecules. By calculating how often and with what force a grain of pollen should collide with a water molecule and comparing this to the observed motions, Einstein was able to estimate the size and mass of a water molecule, and thus the sizes and masses of the hydrogen and oxygen atoms that made up the molecule.

A t an engaging internet site hosted by the American Institute of Physics, you can listen to the voice of Joseph John Thomson talking about his discovery of electrons in 1897. Electrons were the first attack on the atom. At the time of the recording, in 1934, Thomson was seventy-eight years old and had been for many years the Cavendish Professor of Experimental Physics at Cambridge University. The recording crackles with static, but the words are unmistakable: "Could anything at first sight seem more impractical than a body which is so small that its mass is an insignificant fraction of the mass of an atom of hydrogen?" Impractical indeed! But practicality is beside the point here. We're talking about a revolution of ideas, a bombing of the palace of Unity and Indivisibility. A photograph of Thomson at the time shows a deadly serious gentleman, balding, with spectacles and a thick walrus moustache, hands tightly clasped, starched white collar, staring unflinchingly into the camera as if he were looking two thousand years of history in the eye without apologies. "It was coming sooner or later," his gaze seems to say. "So buck up and take it like an adult."

Thomson made his discovery by measuring the paths of electrically charged particles as they were deflected by electric and magnetic forces. First, he and others had to develop good "vacuum pumps" for removing the air in the glass tubes through which the particles moved. Molecules of air interfere with the delicate trajectories of the tiny particles under study. I have a great deal of respect for vacuum pumps. I used them myself during my short-lived encounter with experimental physics as a university student. A vacuum pump, when working properly, starts out with a coarse, grating sound, like the chug of a locomotive, then graduates to a clicking whine, rising in pitch, and ends with a smooth hum when a good vacuum has been attained. When the vacuum is incomplete, the pump never gets past the chugging locomotive stage.

The amount of deflection of a charged particle in a good vacuum indicates the ratio of its electrical charge to its mass. From previous experiments, Thomson and others already knew that particular ratio for hydrogen atoms, the lightest of all atoms. What Thomson found was that these other particles, the electrons—which he called "corpuscles" and which could be created by heating a piece of metal—had a ratio roughly 1,800 times larger than that of hydrogen atoms. Assuming the same electrical charge, the mass was then inferred to be 1,800 times smaller. Evidently, these things were really tiny compared to atoms (although the size of the latter was not known until 1905, as mentioned above). The atom was not the smallest unit of matter.

While Thomson was discovering the electron in England, Antoine Henri Becquerel and Marie Skłodowska Curie were discovering the disintegration of atoms in France, what Madame Curie called "radioactivity." Becquerel believed that the mysterious radiations recently observed to emanate from uranium, the so-called X-rays, were the result of the absorption of sunlight. The uranium X-rays, in turn, could be detected by nearby photographic plates. When Becquerel did his experiment, on February 26, 1896, Paris was cloudy. His uranium did not receive any energizing sunlight. On a whim, he decided to develop his photographic plates anyway. To his surprise, the photographic plates were strongly exposed, showing that the uranium emitted some kind of radiation on its own, without needing to be powered by the Sun. Later experiments by Becquerel showed that the radiations were electrically charged particles of some kind because they were deflected by magnetic fields, as were Thomson's electrons. After the discoveries of Becquerel, Madame Curie did further studies of uranium rays and found that the uranium atoms were hurling out tiny pieces of themselves. A year later, Curie found the same atomic disintegrations with another element, radium. The indivisible atom was, after all, divisible. And what lay inside? No one knew. The bottom of the universe had fallen out.

Here is the reaction of historian Henry Adams in 1903 to these disturbing developments:

As history unveiled itself in the new order, man's mind behaved like a young pearl oyster, secreting its universe to suit its conditions until it had built up a shell of nacre that embodied its notions of the perfect... He sacrificed millions of lives to acquire his unity, but he achieved it, and justly thought it a work of art.

"One God, one Law, one Element" [Adams quoting Tennyson]

Suddenly, in 1900, science raised its head and denied . . . the man of science must have been sleepy indeed who did not jump from his chair like a scared dog when, in 1898, Mme. Curie threw on his desk the metaphysical bomb she called radium.

With his new corpuscles in hand, Professor Thomson proposed what became called the "plum pudding" model of the atom: a tiny ball filled uniformly with a "pudding" of positive electrical charge, into which were sprinkled the negatively charged electrons. You needed the positively charged pudding to balance out the negatively charged electrons, since it was known that most atoms are electrically neutral.

Fifteen years later, the great physicist from New Zealand, Ernest Rutherford, and his assistants found that the atom was not a pudding at all. It was more like a peach. A hard nut resided at its center, containing all of the positive charge and nearly all of the mass. The new particles residing within that hard central nut were called protons and neutrons. Protons have positive electrical charge, neutrons have no charge. This peach picture emerged after Rutherford's team fired subatomic particles at a thin sheet of atoms. Some of the particles veered off at large angles, as if they had hit something hard, a hard nut in the atom. With pudding, the deflections should have been small. "It was quite the most incredible event that had ever happened to me in my life," boomed Rutherford. "It was almost as incredible as if you fired a 15-inch shell at a piece of tissue paper and it came back and hit you." The hard nut at the center of each atom, the "atomic nucleus," is a hundred thousand times smaller than the atom as a whole. To use an analogy, if an atom were the size of Fenway Park, the home stadium of the Red Sox in Boston, its dense central nucleus would be the size of a mustard seed, with the electrons gracefully orbiting in the outer bleachers. In fact, 99.9999999999999 percent of the volume of an atom is empty space, except for the haze of nearly weightless electrons. Since we and everything else are made of atoms, it is literally a fact that we are mostly empty space. That vast emptiness is perhaps the most unsettling consequence of dividing the indivisible.

Eventually, Rutherford's protons and neutrons, at the center of the atom, would themselves be found to consist of even smaller particles called quarks.

Were we falling and falling without end? Were there unlimited infinities on all sides of us, both bigger and smaller, as Pascal believed? It is an unpleasant sensation. I am reminded of the Escher drawing *Ascending and Descending*, which depicts a line of cloaked men walking around a quadrangle in a medieval castle. The disturbing feature of the picture, achieved through a trick of perspective, is that the walkers are always ascending, marching up a continuously rising staircase, and yet after completing the loop they end up exactly where they began. It is a staircase without beginning or end. It is a staircase that goes nowhere.

Escher made *Ascending and Descending* in 1960, at a time when physicists had recently discovered hundreds of novel subatomic particles in the new "atom smashers" and in high-energy radiations from space. The field of research into elementary particles and forces was thrown into chaos. In addition to the electrons and protons and neutrons, there were now delta particles and lambda particles, sigmas and xis, omegas, pions, kaons, rhos, and more. When the Greek alphabet was exhausted, the confounded physicists resorted to using Latin letters. Some of these new subatomic particles had total lifetimes, from the moment they were created to the moment they disappeared, of a mere 10^{-21} seconds, or 0.000000000000000000001 seconds. Before, even with the sacred atom fractured, there had been some kind of order. There had been only the electrons and protons and neutrons. But now—this howling zoo. There seemed to be no fundamental particles, no bottom to the infinite spiral down, no organizing principles.

Then quarks were discovered in the late 1960s. Temporarily, the plummeting stopped. Each of the hundreds of new particles could be understood as a particular combination of a half-dozen basic quarks. Quarks offered a new system for organizing the subatomic zoo. Quarks were the new protons and neutrons, which, in turn, had been the new atoms. I once asked physicist Jerry Friedman, co-discoverer of quarks, whether he thought that the quark was the end of the line, the smallest unit of matter. "Probably," he answered. He gave reasons. But he hesitated. "I could be surprised," he said with a grin. "There are always surprises in science." Surprises in science are good things, and bad.

The philosophers of ancient Greece developed a terrifying view of the world called Zeno's Paradox. Suppose you want to walk 15 feet across a room. Before you travel that distance of 15 feet, however, you must go halfway, which is

7.5 feet. And before you go that 7.5 feet, you must travel half of that distance, 3.75 feet. And before you go that 3.75 feet... And so on. In their minds, the philosophers kept chopping space into halves, into smaller and smaller dimensions ad infinitum, as did Pascal centuries later. The indivisible was pitted against the divisible. The ultimate conclusion of this intellectual exercise is that you cannot cross the room. In fact, you cannot move even an inch. You are frozen in a metaphysical conundrum. You are trapped by the infinity of the small.

When scientists and mathematicians talk about infinity, they are usually imagining a sequence of bigger and bigger spaces and numbers. But infinity can go in the other direction as well. Jerry Friedman, physicist rather than philosopher, is more hopeful. He thinks that the quarks may be the end of the line.

Other physicists disagree. In the last forty years, physicists have proposed objects far smaller than quarks, called "strings." Instead of being point particles, like electrons, strings are extremely tiny one-dimensional "strings" of energy. Their sizes would be the Planck length, where gravity and quantum physics are joined. (See the earlier chapter "Between Nothingness and Infinity.") An important property of strings is that they occupy a space of nine or ten dimensions, instead of the familiar three. In our world of tables and trees, we would not be aware of the additional dimensions because they are curled up into ultra-tiny loops. In the same manner, a garden hose appears as a line when seen at a distance.

Strings were originally proposed as a theory of the strong nuclear force. In more recent years, they have been hypothesized as part of a theory of quantum gravity—that is, Einstein's theory of gravity, general relativity, revised to include quantum physics. At present, no one knows how to test string theory or even whether it can be tested—the sizes are so tiny. Although the mathematics of the theory is beautiful and, in fact, the theory may be the only path to quantum gravity, some physicists have abandoned the theory. For one thing, it may be impossible to test. For another, it has turned out that there are many, many different versions of string theory, each with different outcomes and each corresponding, possibly, to a different universe, with different properties. In that case, our universe would be just one random cosmos, a throw of the dice—defeating the long-standing hope of physicists to show that our universe must necessarily be the way it is and no other way, given a small number of "first principles," in the same way that a crossword puzzle has only one solution.

Regardless of whether strings actually exist, we know that space and time lose their meaning at the Planck size, as discussed in "Between Nothingness and Infinity." We cannot find smaller "particles" beyond Planck; we cannot divide space into smaller elements beyond Planck. It took two thousand years to measure the size of the hypothesized atom. In 1899, Max Planck hypothesized the "Planck length" as the unique length formed by combining his newfound quantum constant with the speed of light and Newton's gravitational constant. Will it be another two thousand years before we can test the existence of strings?

Modern Prometheus

I am by birth a Genevese, and my family is one of the most distinguished of that republic." So begin the reminiscences of Victor Frankenstein in Mary Shelley's famous novel. While at university, during a lightning storm, young Victor sees a stream of fire emerge from a beautiful old oak tree. He becomes enamored of all things scientific and proceeds to study electricity, biology, chemistry, and the new science of galvanism. "One of the phenomena which had peculiarly attracted my attention," recalls Victor years later, "was the structure of the human frame, and, indeed, any animal imbued with life. Whence, I often asked myself, did the principle of life proceed? It was a bold question, and one which has ever been considered as a mystery." After days and nights of laborious experiments, Victor succeeds in discovering how to bring lifeless matter to life. Almost immediately, he decides that he is not satisfied with the bare secret of life but wants to create a human being, with all the intricacies of fibers, muscles, veins—and a brain.

What was the secret that Victor uncovered? For centuries, human beings have puzzled over the mystery of life. What makes an odd hodgepodge of molecules organize itself into living cells, which pulsate and squirm, feed on their surroundings, and then reproduce? Each of us emerged from the cells of our parents, who emerged from their parents, who emerged from their parents, back and back through the dark halls of time. We accept that astounding descent as a given. But how did it start? Surely, that beginning, the origin of life on our planet, and perhaps the origin of life in the entire cosmos, has a significance akin to the origin of the universe itself, borne from the nothingness from which came all matter and energy.

The great biologist Louis Pasteur claimed that life could come only from previous life: *Omne vivum ex vivo.* However, few modern biologists believe that life existed in the early days of our primordial planet, a seething ball of chemicals freshly cooked in the cauldrons of a primordial star. How did it start? Was it an inevitable result of zillions of collisions of atoms, likely to happen on other planets with Earth-like conditions? Or was it a unique occurrence, a one time event? And can physics and chemistry and biology ever give definitive answers to such questions?

Besides these profound scientific questions of origins, there is the philosophical and theological question of the *materiality* of life. Put your finger under a microscope, and you will see cells. Red blood cells, for example, look like red dimpled disks. Examine these cells with a higher-powered microscope, and you'll see tiny hexagons, the molecules of hemoglobin. An even higher power microscope reveals intricate filigrees of oxygen and hydrogen atoms, carbon and nitrogen atoms clustered around an atom of iron. Is that what we are? Is that *all* that we are?

Until recent times, biologists divided into two camps on the question of life.

The so-called mechanists believe that a living creature is just so many atoms and molecules, microscopic pulleys and levers, chemicals and currents—all subject to the laws of chemistry and physics and biology. For that camp, the question of origins amounts to the structure and behavior of atoms and simple molecules, and the energies available in the primitive Earth. Vitalists, on the other hand, argue that there is a special quality of life—some immaterial or spiritual or transcendent force—that enables a jumble of tissues and chemicals to vibrate with life. That transcendent force would be beyond physical analysis or explanation. Some call it the soul. The ancient Greeks called it *pneuma*, meaning "breath" or "wind." Judaism, Christianity, and Islam all hold that the breath of the soul can be imparted only by God.

Modern biologists are mechanists. In fact, an entire interdisciplinary field called synthetic biology is concerned with manufacturing and manipulating components of living systems—aided in part by the discovery of the structure of DNA in the early 1950s and the beginnings of molecular biology. Some synthetic biologists are reprogramming the DNA of microorganisms to produce drugs and batteries and new engineering devices. Others want to understand how life originated on Earth. Still others are attempting to create new forms of life from prior living organisms. Or life from completely nonliving material.

It is a young field. In the 1950s, chemists showed that electrical discharges (lightning) in a mixture of gases thought to represent the ancient atmosphere could produce amino acids, the building blocks of proteins. The first creation of a synthetic cell occurred in the late 1950s and early 1960s. The first hybrid gene, achieved by splicing together the genes of two different organisms, occurred in the early 1970s. The first synthesis of a complete set of genes from their chemical parts and injection into a host cell occurred in 2010. As important as they are, none of these accomplishments comes close to the creation of life from nonlife. However, given the historical momentum of science and the fortitude of the scientists involved, that result is probably only a matter of time. The first human-made life-form, created from scratch, will almost certainly be a single cell with a single gene, far simpler than a bacterium. But that will be a major advance.

Such a result would be the ultimate triumph of the mechanist view. Yet the idea that we may be nothing but material atoms and molecules deeply disturbs many of us. Putting aside for the moment theological considerations, the feeling of self-hood, of thinking and emotion, of self-awareness, of "I-ness" is so overwhelming, so absolutely unique, so impossible to explain, that it seems incomprehensible such a sensation could be rooted completely in material atoms and molecules. It seems impossible that we, and other living beings, could be nothing but material. Yet that is the axiom of the synthetic biologists, who are embarked on a project to create life from nonlife.

If they succeed, their success will reopen many deep questions. At the same time, the ability to create life from nonlife may represent the ultimate freedom of a living being. Not that we will have escaped the laws of nature. But we will have escaped the cosmic decree that living matter emerges from prior living matter in an inevitable chain, unknowing and autonomous, most of it utterly insensible but even the sentient organisms ignorant of the origins of their exquisite bodily machinery. Sometime in childhood, we become aware of ourselves as separate from the surrounding world, as conscious and thinking beings. But we do not remember our birth, or what came before. We understand only a fraction of the trillions of chemical and electrical processes taking place every moment under our skin. We do not know how and why the marvel of our lives, or any life, occurs. We can only accept what is given. If the synthetic biologists succeed in creating life from nonlife, we will be a rare substance in the universe that not only is aware of itself but also understands the secrets of its being.

It is quite possible that the first creation of a living cell from scratch will occur in the laboratory of Jack Szostak, professor of genetics at the Harvard Medical School and professor of chemistry and chemical biology at the Massachusetts General Hospital. Professor Szostak was born in the early 1950s, just at the time that Rosalind Franklin and Francis Crick and

James Watson were making their momentous discoveries about DNA. Szostak grew up in various cities in Germany and Canada as his father, an aeronautical engineer with the Royal Canadian Air Force, was transferred from one posting to the next. For his early fascination with science, Szostak credits his engineer father, who built a basement lab for his son. "The experiments I conducted there often made use of remarkably dangerous chemicals that my mother was able to bring home from the company where she worked," recalls Szostak. He also credits his father with his decision at an early age to become an academic. "My father was often unhappy with his job, chafing at both his superiors and his subordinates. This I am sure made me seek out the academic life for its more egalitarian aspects. I have never felt like I worked for a boss or had employees who worked for me, just colleagues who like me were interested in learning more about the world around us."

In 1968, at the age of fifteen, young Szostak began his undergraduate studies at McGill. His particular engagement with biology was sparked by a summer program for undergraduates at the Jackson Laboratories on Mount Desert Island, off the coast of Maine, where he analyzed the thyroid hormones of mice. In the early 1970s, Szostak began his graduate studies at Cornell. There, he worked with the DNA of yeast. Over the next decade and a half, that work deepened and spread, culminating in Szostak's discovery of how the vulnerable ends of yeast chromosomes, and indeed all chromosomes, are protected by molecules called "telomeres"—work for which he won the 2009 Nobel Prize in Physiology or Medicine (shared with Elizabeth Blackburn and Carol Greider).

By the mid to late 1980s, the field of yeast biology was getting too congested for Szostak. "I had a growing feeling that my work in yeast was becoming less significant, in the sense that other people would inevitably end up doing the same experiments we were doing in a few months or years at the most," he recalls. From the beginning of his career, Szostak had always tried to avoid direct competition with other scientists. So, still in his thirties, with his Nobel Prize–winning work already under his belt, he began shifting his focus to RNA (ribonucleic acid), a molecule very similar to DNA and thought to be its ancestor in the evolution of life. Since then, Szostak and the researchers in his lab have been at the forefront of the creation of life from nonlife. Some of their major achievements include the creation of cell membranes from simple chemicals and a demonstration of how these membranes could grow and divide under simple chemical and physical processes, and a partial understanding of how RNA can be replicated within a primitive surrounding membrane.

Biologists are not in complete agreement about when to declare a particular smidgeon of matter "alive." In general, the requirements include some kind of surrounding membrane (what Szostak calls a "compartment") to separate the organism from the outside world and to confine the most critical molecules in close proximity, the ability to utilize energy sources, the ability to grow, the ability to reproduce, and the ability to evolve. In a 2001 paper in the prominent journal *Nature*, Szostak and colleagues identified four vital ingredients of a minimal living cell: a compartment, an embedded molecule like RNA or DNA that is able to replicate, a means for that replication, and some kind of interaction between the compartment wall and the replicating molecule so that they can help each other in response to the forces of Darwinian evolution. What distinguishes Szostak's work in this field from the work of many other synthetic biologists is that Szostak wants to create a living cell from scratch, using only the simple molecules present in the primordial Earth, what he calls "prebiotic" molecules. By contrast, most other labs start with complex molecules that have been snatched from existing life-forms and have already had the benefit of natural selection and evolution over hundreds of millions of years.

Although ambitious and laser focused, Szostak is exceptionally modest about his achievements. Near the beginning of his autobiographical statement upon receiving the Nobel in 2009, he says: "Although I have had some degree of success as a scientist, it is hard to say precisely why." He is also exceptionally generous in giving credit and support to others. "One of the delights of the world of science," he says, "is that it is filled with people of good will who are more than happy to assist a student or colleague by teaching a technique or discussing a problem." He refers to his second graduate student, Andrew Murray, as "a brilliant and energetic student who was fun to talk with about any conceivable experiment." About another of his students, he recalls, "I had the good fortune to 'inherit' one of [Harvard chemist Jeremy Knowles's] graduate students, Jon Lorsch, who migrated to my lab and did outstanding work on ribozyme selections and mechanistic enzymology." A photograph of Professor Szostak with his students in a quite plain-looking room at the Harvard Medical School shows twenty smiling young people, some standing, some kneeling, most wearing jeans. In the middle of this happy family is the professor. By his humble demeanor, he is one of them.

I visited Professor Szostak in his office and lab in July 2019. His office, on the fourth floor of the Richard B. Simches Research Center of the Massachusetts General Hospital, is a small room barely large enough to hold a small couch, a small table, a small desk piled high with papers and documents, and a bookshelf with biology books and bound volumes of his students' theses. When I met him, he was wearing a sweat-stained blue linen shirt and rumpled khaki pants loosely drooping from his waist. He has thinning hair and wears glasses. He speaks in a soft, almost hesitant voice, clearly passionate about his work while at the same time not endowing any sentence with the slightest exaggeration or presumption of importance. "People get all tied up in knots about defining life," he told me. "That doesn't help us. I care about the process and pathway from simple to complex. Where along the way you draw the line and call the thing 'alive'—different people draw the line at different places. If it can start evolving, I would call it alive." Evolution and natural selection are powerful driving forces. Szostak points out that any biological molecule will naturally undergo mutations, some positive and some negative. Given the right chemical environment, evolution then happens automatically. "Once you have one element that has an advantage, there is a huge pressure to drive replication... When somebody figures it out [how life started on Earth], it's got to be a bunch of simple things." He smacked his hand against his head in a eureka gesture. "It happened all by itself on primitive Earth. It can't be that hard."

In 2003, Szostak and his colleagues demonstrated that a common mineral clay called montmorillonite, formed from volcanic ash and used in cat litter today, could accelerate the assembly of cell "compartments" needed for life using only the simple molecules available in the primordial Earth. Montmorillonite seems to be an extraordinary catalyst. It was already known that it could help assemble RNA molecules from their basic building blocks. Now, Szostak and his colleagues found that simple molecules called fatty acids, when placed in contact with the clay, bond together to form membranes. The membranes then automatically close up and assemble tiny fluid-filled sacs, or compartments, which could possibly contain replicating molecules like RNA or DNA. Furthermore, in the presence of the clay, these microscopic sacs grow all by themselves by incorporating other fatty acids. Evidently, the surface of the clay has special geometrical and chemical properties that catalyze these reactions. Szostak and his colleagues also showed that passing the tiny sacs through a material with small pores would cause them to divide, in a sense "reproducing." Thus, he had demonstrated creation, growth, and reproduction of a cell compartment.

Almost immediately after his paper was published in *Science*, it was widely popularized in the press. For example, *The New York Times* published an article about the work titled "How Did Life Begin?" and *Scientific American* published an article titled "Clay Could Have Encouraged First Cells to Form."

Szostak wanted to tell me a story about this discovery. After it was picked up by the news media, he received a "flood of emails" from fundamentalists saying they were pleased he had proven that God can create life from clay, just as mentioned in the sacred books. "I am not religious myself," he said and smiled at the irony. "I hope that when we succeed, it will eventually seep into the culture that the creation of life is totally natural, and we don't need to invoke anything magical or supernatural... What I don't get is how religious people can say that they know how God did that."

While we talked, several of Professor Szostak's students and colleagues quietly worked in the lab just outside his office. His current research group consists of sixteen students and postdocs. The major part of the lab occupies part of a large room housing a dozen or so long shelves cluttered with various bottles and chemicals. Below the shelves are worktables. On one worktable I saw a computer screen, an open notebook and pen, and several Post-it notes stuck to the wall and shelf. Adjoining this large room are a few smaller rooms with mass spectrometers (which measure the ratio of mass to electrical charge of tiny particles and thus help identify them); centrifuges; an oxygen-free zone, contained within an airtight hood and used to simulate the oxygen-free atmosphere of the early Earth; and a sophisticated nuclear magnetic resonance (NMR) machine to measure the structure of molecules. As I stood gaping at the NMR machine, Szostak mentioned that he would like to have two, as a backup when one of them is temporarily down.

Szostak and many other biologists who study the origin of life subscribe to a view called the "RNA world." This concept, first proposed by biologist and biophysicist Alexander Rich in 1962, holds that the first replicating molecule in the early history of Earth was not DNA but RNA. The two molecules are chemical cousins. They differ in a few ways. In modern cells, most DNA is a double-stranded helix while most RNA is single-stranded; one of the four letters of the genetic alphabet used by the two molecules is different; and the backbones of the two molecules incorporate slightly different sugar molecules. (The sugar molecule found in DNA derives from the simpler sugar molecule in RNA, another reason why many biologists believe that RNA came first.) Both RNA and DNA store information for the reproduction of the organism. Unlike DNA, RNA has other duties in the cell. It reads the information on the DNA molecule and then carries that information to another part of the cell where proteins are made.

The RNA world hypothesis got a big boost in the early 1980s when biologists Thomas Cech and Sidney Altman independently showed that RNA was not simply a passive messenger of information but could catalyze reactions and help create molecules on its own. This discovery solved a long-standing chicken-and-egg-type conundrum: certain proteins are needed to make DNA, but DNA is needed to make those proteins. RNA could do both: store genetic information for the cell and also rebuild itself. RNA could be both the carrier of the map and the mapmaker.

Being single-stranded, RNA is more subject to attack and degradation by outside chemicals. It is not as stable as DNA. Over time, in the process of Darwinian evolution, RNA would have been replaced by DNA as the principal bank of genetic information. But in the beginning, according to RNA world, RNA might have been the principal molecule of replication.

Szostak believes that a primitive cell might not need much more in its innards than a strand of RNA and some simple chemicals to serve as raw construction materials. How that construction occurs is not yet understood—a major obstacle for understanding how to create life from nonlife. "In my view, the critical problem right now is to understand the chemistry that enabled the first mode of RNA replication," says Szostak. In other words, exactly how does the replication molecule, carrying all the blueprints for the cell, replicate itself? Szostak says that it is easy to replicate RNA using protein enzymes (catalysts) and other complex molecules that have been developed over millions of years of evolution. But he wants to know how life began. He is trying to show how RNA replication could have happened on the primitive Earth, with only the simple molecules then in existence. "Our approach has a ways to go: so far, we can copy short stretches of an RNA template to generate a complementary strand in the form of an RNA double helix. However, our ability to copy RNA is limited to very short lengths, and we cannot yet do multiple cycles of copying, in other words, copy the copies. Indefinite replication within protocells is our goal, because we think that with replicating RNA inside replicating vesicles [membrane compartments], we would have a system capable of evolving in a Darwinian sense."

The study of how life began on Earth, and the related attempt to create life from nonlife in the lab, raise all kinds of philosophical, theological, ethical, and social issues. Many of these issues have been anticipated in science fiction, in academia, and in religious conferences and institutions. But with the successes of Szostak and other synthetic biologists, these matters are receiving new attention.

In one episode of *Star Trek: The Next Generation*, Commander Data fractures a part of himself and stares at the bare tangle of wires and computer chips protruding from his wrist. Although Data is a machine, viewers have come to regard him as human. He looks human. He acts toward other characters with compassion and sweetness. He appears to know right from wrong. Something unsettles us about this scene, not so much because Data is hurt, but because he, and we, suddenly see inside his mechanism. The secret of his being hangs open in the air. All the complexities of his bodily actions and thoughts, the subtle depths of his feelings, the seemingly infinite mysteries of a living being, have been graphically reduced to so many amperes of current flowing through these protruding wires, to particular patterns of zeros and ones within these computer components. We feel affronted. We feel some kind of violation of the natural order of things.

In our age of rampant technological advance—with small boxes that can transmit our words and images over vast distances through space, with other devices that improve our hearing and sight, with drugs that alter our thoughts and personalities, all of it human-made—the lines between the "natural" and the "unnatural" have become blurred. One might argue that since we human beings are "natural" and our brains and their capacities evolved "naturally," then anything we make is "natural." Others disagree. What are the differences, if any, between an organism created in Jack Szostak's lab and an organism found in the moist soil under a rock?

Micah Greenstein, a prominent rabbi in Memphis, Tennessee, says unequivocally that an organism created in a lab would not have a soul. "Soul is the life force within all living beings which cannot be quantified," says Rabbi Greenstein. "It is the animating feature of all life-forms. Dogs have souls. They cry and offer compassion and love like humans. Human souls are gifted with the ability to take care of other life-forms and the planet itself. Were we able to construct a new life, I believe no process could 'breathe' into a proto-human or proto-dog the soul of which I speak. Call it personality, call it the unique signature of each human being. In a beautiful midrash, the rabbis speak of the difference between a coin creator and God. A coin maker puts the same mark on each shekel and they are exactly the same. God breathes 'soul' into every human, the same gift of spirit in every human being, yet no two people are exactly alike. Each is a unique signature of the One."

A related concern for theists is that the human creation of living organisms seemingly transgresses into territory and knowledge reserved for God. This concern has a long history. In Milton's *Paradise Lost* (1667)—written in the age of Newton and the beginnings of modern science—when Adam questions the angel Raphael about celestial mechanics, Raphael offers some vague hints and then says that "the rest / From Man or Angel the great Architect / Did wisely to conceal, and not divulge / His secrets to be scann'd by them who ought / Rather admire." In 1996, when the British embryologist Dr. Ian Wilmut cloned a lamb (named Dolly) from an adult sheep's cell, it was undoubtedly a success for science, but ethical and theological alarms sounded all over the world. The sheep's human manipulators were described by *The New York Times* as having "pried open one of the most forbidden—and tantalizing—doors of modern life." Cloning is a complex issue, and one has to distinguish between therapeutic cloning (to treat illnesses) and reproductive cloning (to produce new organisms). On the subject of human cloning, most religious groups are adamantly opposed, or highly suspicious. But even with Dolly, many were disturbed when the achievement was announced. Dozens of articles had the phrase "playing God" in the title. Even today, according to a recent Gallup survey, 66 percent of Americans think it is "morally wrong" to clone animals, while 31 percent say it is "morally acceptable."

Ruth Faden, professor of biomedical ethics at Johns Hopkins University and founder of the Berman Institute of Bioethics at that university, frames the question of human-made organisms in terms of their "moral status." An entity's "moral status" determines its rights and value and sets parameters for how we human beings are morally obligated to treat it. The term has gained currency in the twentieth century with the abortion debate and questions about whether a human embryo has moral status. Professor Faden discussed the issue with me specifically in relation to synthetic biology. "Does it matter how life is formed?" she said. "Certainly it matters scientifically. But does it matter for the entity that is created? Should it be treated any differently from something we just found under a rock? For some people, in the ethical domain there is a big line between organic and inorganic entities. Living matter has value in ways that inorganic matter does not. There is a huge debate over who has moral status." Faden says that for people of faith, "we may be redefining where the spark of life is." She adds that "for many nonreligious people, it is not the soul that matters but sentience," suggesting that the dividing line between deserving some level of moral status and not is whether the organism has feelings and consciousness. Consciousness and self-awareness are not easy to pin down, but wherever one draws the line it is likely that at some point in the future biologists will have the ability to create a sentient living being.

At least for the creators and writers of *Star Trek*, Commander Data has moral status. Did the creature brought to life by Victor Frankenstein have moral status? Does a computer with the ability to learn and speak have moral status? In Rabbi Greenstein's view, none of these entities would have a soul, but some of them would be sentient, however one defines that word. The Buddhist monk Yos Hut Khemacaro, who played a major role in rebuilding the monkhood in Cambodia in the 1980s and 1990s, told me that Buddhists, who do not believe in a soul, have no problem with human-made life-forms and would accord them "morality, value, and dignity if we see the same characteristics as 'natural' living beings."

Some observers place synthetic biology within the larger context of our exponentially advancing technology in general and the need for restraint. Richard Hayes, a social and political advocate and former executive director of the Center for Genetics and Society in Berkeley, says, "We're at or very near a cusp in the history of humankind. It's no longer a question of the pros and cons of this one new technology or that particular application. I believe we need to take a huge, deep breath, take a big step back and give ourselves the time and space within which to assess where we are, how we got here and where we want to go, along the entire set of social, political and technological dimensions. We need to draw lines. If we allow scientists to create a single living cell, for example one very effective at pulling CO_2 from the atmosphere, why not allow creation of two-celled organisms that do so even more effectively, or of 200- or 2,000-celled organisms that remove pollutants from the ocean? Why not fish-like or rat-like organisms that possess certain human cognitive abilities and can be trained for many useful purposes? And if that's OK, why not allow the creation of human-ape hybrids to perform even more useful tasks?"

Certainly, there are safety issues. In the early 1970s, Paul Berg at Stanford University produced a hybrid loop of DNA containing DNA from two different organisms, a virus called SV40 and the common bacterium *E. coli*. Berg was planning on inserting this human-made recombinant DNA back into *E. coli*. When concerns were raised about the unforeseen consequences of creating an organism never before seen in nature, Berg suspended his experiments. At that point, the U.S. National Academy of Sciences appointed a committee to study the safety issues of recombinant DNA research. After the committee published its report in 1974, the scientists recommended a worldwide deferment on certain kinds of recombinant DNA research until the risks were better understood. "There is a serious concern that some of the artificial recombinant DNA molecules could prove biologically hazardous," wrote the committee. Today, forty-five years later and with guidelines in place, recombinant DNA technologies have been enormously useful in producing new vaccines, protein therapies such as human insulin, blood-clotting factors, and gene therapy.

A more recent development occurred in 2010, when J. Craig Venter and colleagues created a set of genes that were a variant of already existing bacterial genes and then inserted them into a bacterium that had had its own DNA removed. The synthetic genes then took over the bacterium. The accomplishment triggered a presidential investigative commission, under President Obama. In its report, titled "The Ethics of Synthetic Biology and Emerging Technologies," the commission wrote: "The Venter Institute's research and synthetic biology are in the early stages of a new direction in a long continuum of research in biology and genetics. The announcement last May [of Venter's achievement], although extraordinary in many ways, does not amount to creating life as either a scientific or a moral matter... In order to provide benefits to human conditions and the environment, the Commission thinks it imprudent either to declare a moratorium on synthetic biology until all risks can be determined and mitigated, or to simply 'let science rip,' regardless of the likely risks... The Commission instead proposes a middle ground—an ongoing system of prudent vigilance that carefully monitors, identifies, and mitigates potential and realized harms over time."

In 1981, several years before his death, the great theoretical physicist Richard Feynman did an interview for the BBC television program *Horizon,* in which he was asked a question about his Nobel Prize. Feynman's reply: "I don't see that it makes any point that someone in the Swedish Academy decides that this work is noble enough to receive a prize— I've already got the prize. The prize is the pleasure of finding the thing out, the kick in the discovery, the observation that other people use it [my work], those are the real things..." Jack Szostak, too, has a Nobel. And he is undoubtedly aware of many of the theological, ethical, and philosophical aspects of his work. He is also aware of the medical and business opportunities of synthetic biology in general. (In the 1990s, he and two colleagues founded a start-up biotechnology company to produce new kinds of proteins. "Although the company was not a business success, it was a very interesting and educational experience," he recalls.) But what drives Szostak and many other basic scientists—what keeps them up late at night in the lab or at their work desk, so that they can think of nothing else, sometimes to the neglect of their family and friends—is what drove Feynman: "the pleasure of finding the thing out." How did life begin on our planet? What did the first replicating cells look like? How do we create a living thing from nonliving material, life from nonlife—a squirming, growing, evolving, reproducing thing from simple chemicals? Few questions could be more profound. Yet it is not only the profundity of the questions. It is the primal pleasure of finding things out, and the incomparable thrill of being the first person to understand something about nature.

As I talked with Professor Szostak about his research, despite the quiet of his voice I could hear the passion. In his autobiography, he wrote these words about his work to create primitive replicating molecules (which he calls modified nucleic acids or genetic polymers): "It is thrilling to me to see people in my lab developing new approaches to the synthesis of modified nucleic acids, but the suspense is almost unbearable as we await the results of template-directed polymerization experiments. From our current vantage point, it is not clear whether there will be many solutions to the problem of chemically replicating genetic polymers, or just one, or none, but in any case it is an exciting quest."

There is one significant way in which Szostak's pleasure in science differs from Feynman's. As a theoretical physicist, Feynman worked alone. By contrast, Szostak and most biologists today work in groups, surrounded by a team of graduate students, postdocs, and other colleagues. It is a more social enterprise. And that companionship provides Szostak, and other higher life-forms, with an additional pleasure. Near the end of my visit with Professor Szostak, he summed up the last decade: "What I love doing the most is talking to other colleagues, students, and postdocs. One of the best things about having a lab is helping young people develop."

MIND

MIND

One Hundred Billion

I've always been struck by the fact that the number of neurons in our brain is about equal to the number of stars in a galaxy: one hundred billion. The first reflects the architecture of one unit of consciousness, a mind, and the second one unit of glowing cosmic matter as seen by a giant being. Perhaps we should not make much of this coincidence. Still, it reminds us of our place in the cosmos, just as Copernicus and Darwin reminded us and reconfigured that place.

Not only are we cosmic material. We are the *precise* material made in stars. Our atoms, our particular atoms, one by one, were forged in the nuclear reactions of stars, then hurled out into space in the explosion of those stars, to swirl and condense millions of years later into planets and ultimately into single-celled organisms and ultimately into us human beings. We are literally part of the cosmos. Contrary to widespread belief, there are not two kinds of material in the cosmos—inanimate material, like rocks and water and planets and stars, plus a second kind of material, the animate, endowed with some supernatural, transcendent essence. There is only a single kind of material, made of atoms. Rocks, water, air, trees, human beings—all are constructed of the same atoms.

Still, it is stunning that a mere assemblage of atoms can produce the exquisite sensation of our consciousness, feelings of love and anger, self-awareness and self-reflection, memory, painters and philosophers and scientists. How is that possible? The British philosopher Colin McGinn has argued that we can never understand consciousness because we can never get outside of our minds to do the analysis. We are necessarily trapped within three pounds of moist gray matter, thinking and perceiving within that constraint. Whether McGinn is right or wrong, we certainly must acknowledge that any discussion of the physical cosmos is predicated on our perceptions, our language, the instruments we build. And any discussion of our personal experience with the world must include memory and the vagaries of memory. For us human beings, our minds are necessarily part of our description of reality. We study other animals, plants, nuclear reactions, cell division, DNA, planets, stars. And we ourselves are always necessarily involved with such studies, because we cannot think outside of our minds. Thus it seems natural that the scientist and mathematician Pascal, when contemplating the infinitely small and the infinitely large within the cosmos, included human beings in the same paragraph. But, as I have said before, that infinity is not to be feared. Rather, it is to be embraced. We are part of it.

Many years ago, I took my two-year-old daughter to the ocean for the first time. We parked our car some distance away, beyond sight of the water, and then walked across a broad sandy area, passing sand dunes, shells of crabs, piping plovers that ran and stopped, ran and stopped, ran and stopped. Eventually, we climbed over a sandy hill. And there was the ocean, stretching on and on until it merged with the sky. It was my daughter's first glimpse of infinity. For a moment, her face froze. Then she broke out in a big smile.

Smile

It is a Saturday in March. The man wakes up slowly, reaches over and feels the windowpane, and decides it is warm enough to skip his thermal underwear. He yawns and dresses and goes out for his morning jog. When he comes back, he showers, cooks himself a scrambled egg, and·settles down on the sofa with the *Essays of E. B. White.* Around noon, he rides his bike to the bookstore. He spends a couple of hours there, just poking around the books. Then he pedals back through the little town, past his house, and to the lake.

When the woman woke up this morning, she got out of bed and went immediately to her easel, where she picked up her pastels and set to work on her painting. After an hour, she is satisfied with the light effect and quits to have breakfast. She dresses quickly and walks to a nearby store to buy shutters for her bathroom. At the store, she meets friends and has lunch with them. Afterward, she wants to be alone and drives to the lake. Now, the man and the woman stand on the wooden dock, gazing at the lake and the waves on the water. They haven't noticed each other.

The man turns. And so begins the sequence of events informing him of her. Light reflected from her body instantly enters the pupils of his eyes, at the rate of ten trillion particles of light per second. Once through the pupil of each eye, the light travels through an oval-shaped lens, then through a transparent, jelly-like substance filling up the eyeball, and lands on the retina. Here it is gathered by one hundred million rod and cone cells.

Cells in the path of reflected highlights receive a great deal of light; cells falling in the shadows of the reflected scene receive very little. The woman's lips, for example, are just now glistening in the sunlight, reflecting light of high intensity onto a tiny patch of cells slightly northeast of the back center of the man's retina. The edges around her mouth, on the other hand, are rather dark, so that cells neighboring the northeast patch receive much less light.

Each particle of light ends its journey in the eye upon meeting a retinene molecule, consisting of 20 carbon atoms, 28 hydrogen atoms, and 1 oxygen atom. In its dormant condition, each retinene molecule is attached to a protein molecule and has a twist between the eleventh and fifteenth carbon atoms. But when light strikes it, as is now happening in about 30,000 trillion retinene molecules every second, the molecule straightens out and separates from its protein. After several intermediate steps, it wraps into a twist again, awaiting arrival of a new particle of light. Far less than a thousandth of a second has elapsed since the man saw the woman.

Triggered by the dance of the retinene molecules, the

nerve cells, or neurons, respond. First in the eye and then in the brain. One neuron, for instance, has just gone into action. Protein molecules on its surface suddenly change their shape, blocking the flow of positively charged sodium atoms from the surrounding body fluid. This change in flow of electrically charged atoms produces a change in voltage that shudders through the cell. After a distance of a fraction of an inch, the electrical signal reaches the end of the neuron, altering the release of specific molecules, which migrate a distance of a hundred-thousandth of an inch until they reach the next neuron, passing along the news.

The woman, in fact, holds her hands by her sides and tilts her head at an angle of five and a half degrees. Her hair falls just to her shoulders. This information and much, much more is exactingly encoded by the electrical pulses in the various neurons of the man's eyes.

In another few thousandths of a second, the electrical signals reach the ganglion neurons, which bunch together in the optic nerve at the back of the eye and carry their data to the brain. Here, the impulses race to the primary visual cortex, a highly folded layer of tissue about a tenth of an inch thick and two square inches in area, containing one hundred million neurons in half a dozen layers. The fourth layer receives the input first, does a preliminary analysis, and transfers the information to neurons in other layers. At every stage, each neuron receives signals from a thousand other neurons, combines the signals—some of which cancel one another out—and dispatches the computed result to a thousand-odd other neurons.

After about thirty seconds—after several hundred trillion particles of reflected light have entered the man's eyes and been processed—the woman says hello. Immediately, molecules of air are pushed together, then apart, then together, beginning in her vocal cords and traveling in a spring-like motion to the man's ears. The sound makes the trip from her to him (twenty feet) in a fiftieth of a second.

Within each of his ears, the vibrating air quickly covers the distance to the eardrum. The eardrum, an oval membrane about 0.3 of an inch in diameter and tilted 55 degrees from the floor of the auditory canal, itself begins trembling and transmits its motion to three tiny bones. From there, the vibrations shake the fluid in the cochlea, which spirals snail-like two and a half turns around.

Inside the cochlea, the tones are deciphered. Here, a very thin membrane undulates in step with the sloshing fluid, and through this basilar membrane run tiny filaments of varying thicknesses, like strings on a harp. The woman's voice, from afar, is playing this harp. Her hello begins in the low registers and rises in pitch toward the end. In precise response, the thick filaments in the basilar membrane vibrate first, followed by the thinner ones. Finally, tens of thousands of rod-shaped bodies perched on the basilar membrane convey their particular quiverings to the auditory nerve.

News of the woman's hello, in electrical form, races along the neurons of the auditory nerve and enters the man's brain, through the thalamus, to a specialized region of the cerebral cortex for further processing. Eventually, a large fraction of the hundred billion neurons in the man's brain become involved with computing the visual and auditory data just acquired. Sodium and potassium gates open and close. Electrical currents speed along neuron fibers. Molecules flow from one nerve ending to the next.

All of this is known. What is not known is why, after about a minute, the man walks over to the woman and smiles.

The Anatomy of Attention

Every moment, our brains are bombarded with information, from without and within. The eyes alone convey more than one hundred billion signals to the brain every second. The ears receive another avalanche of sounds. Then there are the internal fragments of thoughts, conscious and unconscious, racing from one neuron to the next. Much of this data is random and meaningless. Indeed, for us to function, much of it must be ignored. But clearly not all. How do our brains select the relevant data? How do we decide to pay attention to the beep of a smoke alarm and ignore the drip of a leaky faucet? How do we become conscious of a certain stimulus, or indeed "conscious" at all?

For decades, psychologists, philosophers, and scientists have debated the process by which we pay attention to things, based on cognitive models of the mind. But in the view of modern scientists, the "mind" is not some nonmaterial and exotic essence separate from the body. All questions about the mind must ultimately be answered by studies of physical cells, explained in terms of the detailed workings of the hundred billion neurons in the brain. At this level, the question is, How do a group of neurons signal to one another and to a cognitive command center that they have something important to say?

"Years ago," the neuroscientist Robert Desimone told me during a recent visit in his office, "we were satisfied to know which areas of the brain light up under various stimuli. Now, we want to know *mechanisms*." Desimone directs the McGovern Institute for Brain Research at the Massachusetts Institute of Technology. Youthful and trim at age sixty-two, he was casually dressed in a blue pin-striped shirt, with only the slightest gray in his hair. On the bookshelf of his tidy office were photographs of his two young children; on the wall was a large watercolor titled *Neural Gardens*, depicting a forest of tangled neurons, their spindly axons and dendrites winding downward like roots in rich soil.

In an article published in the journal *Science* in 2014, Desimone and his colleague Daniel Baldauf reported on an experiment that shed light on the physical mechanism of paying attention. The researchers presented a series of two kinds of images, faces and houses, to their subjects in rapid succession, like passing frames of a movie, and asked them to concentrate on the faces but disregard the houses (or vice versa). The images were "tagged" by flashing them at two different frequencies—a new face image every two-thirds of second and a new house image every half second. By monitoring the frequencies of the electrical activity of the subjects' brains with magnetoencephalography (MEG) and functional magnetic resonance imaging (fMRI), Desimone and Baldauf could determine where in the brain the images were being directed.

The scientists found that even though the two sets of images were presented to the eye almost on top of each other, they were processed by different places in the brain: the face images by a particular region on the surface of the temporal lobe that is known to specialize in face recognition, and the house images by a neighboring but separate group of neurons specializing in place recognition.

Most importantly, Desimone and Baldauf found that the neurons in the two regions behaved differently. When the subjects were told to concentrate on the faces but to disregard the houses, the neurons in the face location fired in synchrony, like a group of people singing in unison, while the neurons in the house location fired like a group of people singing out of synch, each beginning at a random part of the song. And when the subjects concentrated on houses and disregarded the faces, the reverse happened. Furthermore, another part of the brain called the inferior frontal junction, a marble-sized region in the frontal lobe, seemed to orchestrate the chorus of the synchronized neurons, since it fired slightly ahead of them. Evidently, what we perceive as "paying attention" to something originates, at the cellular level, in the synchronized firing of a group of neurons, whose rhythmic electrical activity rises above the background chatter of the vast neuronal crowd. Or, as Desimone once put it, "This synchronized chanting allows the relevant information to be 'heard' more efficiently by other brain regions."

A connection between attention and neuronal synchrony was first hypothesized by Ernst Niebur and Christof Koch twenty years ago. Desimone was one of the first scientists to prove it for particular cases, in 2001. A pioneer in the field, he is quick to mention other leaders, such as John Reynolds of the Salk Institute, who uses a combination of physics, neurophysiology, and computational neural modeling to study how simultaneous objects in the visual field, such as separate highlighted areas in an illuminated grid, compete with one another for attention. Meanwhile, Sabine Kastner of Princeton has recently begun comparing humans with monkeys in their attention to visual tasks; and Columbia's Michael Goldberg has recently shown that, in the process of attention, a particular area of the brain called the lateral parietal area "sums up" visual signals and cognitive signals. In this growing field of neuroscience, Desimone has trained over thirty-five people himself.

I asked Desimone how the conductor of the neuronal chorus, in this case the inferior frontal junction, would know that a particular stimulus should be attended to. In his experiment, the subjects were told to focus their attention on either faces or houses, but what about an unexpected stimulus—say a charging lion or the sudden entrance of a potential romantic partner? "We don't understand the answer to that yet," said Desimone. And how do a bunch of random voices come into synchrony? Can they do so merely by exchanging notes with one another, or do they need an outside director? At the second question, Desimone broke out in a boyish grin and took six small metronomes from his briefcase. He placed them side by side on a wooden board, balanced on two empty lemon soda cans. Then he set the metronomes ticking, out of synch with one another. After a couple of minutes, they were all ticking in synchrony. They had communicated with one another and come into synch solely through the vibrations of the board, without any outside agency. Neurons, of course, use a different method of communication with one another: passing chemical messengers between the hundreds of root-like filaments radiating from each neuron. Desimone's pendulums suggest that some neurons could come into synch on their own, without a conductor. But the question of which neuronal processes are self-organizing and which require a higher-level cognitive director isn't yet understood.

As my visit came to an end, I asked Desimone about the seemingly strange experience of "consciousness," to me the most profound and troubling aspect of human existence. How does a gooey mass of blood, bones, and gelatinous tissue become a sentient being? How does it become aware of itself as a thing separate from its surroundings? How does it develop a self, an ego, an "I"? Without hesitation, Desimone replied that the mystery of consciousness was overrated. "As we learn more about the detailed mechanisms in the brain," he said, "the question of 'What is consciousness?' will fade away into irrelevancy and abstraction." As Desimone sees it, consciousness is just a vague word for the mental experience of attending, which we are slowly dissecting in terms of the electrical and chemical activity of individual neurons. He threw out an analogy. Consider a careening automobile. A person might ask: Where inside that thing is its *motion*? But he would no longer ask that particular question after he understood the engine of the car, the manner in which gasoline is ignited by sparkplugs, the movement of cylinders and gears.

I am a scientist and a materialist myself, but I left Desimone's office feeling somehow bereft. Although I cannot say exactly why, I do not want my thoughts, my emotions, and my sense of self reduced to the electrical tinglings of neurons.

I prefer that at least some parts of my being remain in the shadows of mystery.

Immortality

Early August. I am lying in a hammock and musing on mortality. A hundred years from now, I'll be gone, but many of these spruce and cedars will still be here. The wind going through them will still sound like a distant waterfall. The curve of the land will be the same as it is now. The paths that I wander may still be here, although probably covered with new vegetation. The rocks and ledges on the shore will be here, including a particular ledge I'm quite fond of, shaped like the knuckled back of a large animal. Sometimes, I sit on that ledge and wonder if it will remember me. Even my house might still be here, or at least the concrete posts of its footing, crumbling in the salt air. But eventually, of course, even this land will shift and change and dissolve. Nothing persists in the material world. All of it changes and passes away.

That said, I think that the distinction between life and death may be overrated. I have come to believe that death occurs gradually, through the diminishing of consciousness.

Let me explain. According to the scientific view, we are made of material atoms, and nothing but material atoms. To be precise, the average human being consists of about 7×10^{27} atoms (seven thousand trillion trillion atoms)—65% oxygen, 18% carbon, 10% hydrogen, 3% nitrogen, 1.4% calcium, 1.1% phosphorous, and a smattering of 54 other chemical elements. The totality of our tissues and muscles and organs is composed of these atoms. And, according to the scientific view, there is nothing else. To an alien intelligence, each of us human beings would appear to be an assemblage of atoms, humming with our various electrical and chemical energies. To be sure, it is a special assemblage. A rock does not behave like a person. But the mental sensations we experience as consciousness and thought, according to science, are purely material consequences of the purely material electrical and chemical interactions between neurons, which in turn are simply assemblages of atoms. And when we die, this special assemblage disassembles. The atoms remain, only scattered about.

Particularly special in these considerations is the brain. In the view of science, the brain is where our self-awareness originates, our memories are stored, our elusive ego and "I-ness" are formed. Neuroscientists like Robert Desimone of MIT have studied the brain in great detail. Much is known. Much remains unknown. But the materiality of the organ is not in doubt. There is good evidence that the processing and storage of information is done by the brain cells called neurons. There are about a hundred billion neurons in the average human brain, and each neuron is connected by long filaments to between a thousand and ten thousand other neurons. The electrical and chemical components of these neurons are largely understood.

Despite the known material nature of the brain, the sensation of consciousness—of ego, of "I-ness"—is so powerful and compelling, so fundamental to our being and yet so difficult to describe, that we endow ourselves and other human beings with a mystical quality, some magnificent and nonmaterial essence that blooms far larger than any collection of atoms. To some, that mystical thing is the soul. To some it is the Self. To others, it is consciousness.

The soul, as commonly understood, we cannot discuss scientifically. Not so with consciousness, and the closely related Self. Isn't the experience of consciousness and Self an illusion caused by those trillions of neuronal connections and electrical and chemical flows? If you don't like the word *illusion*, then you can stick with the sensation itself. You can say that what we call the Self is a name we give to the mental sensation of certain electrical and chemical flows in our neurons. That sensation is rooted in the material brain. And I do not mean to diminish the brain in any way by affirming its materiality. The human brain is capable of all of the wondrous feats of imagination and self-reflection and thought that we ascribe to our highest existence. But I do claim that it's all atoms and molecules. If the alien intelligence examined a human being in detail, he/she/it would see fluids flowing, sodium and potassium gates opening and closing as electricity races through nerve cells, acetylcholine molecules migrating between synapses. But he/she/it would not find a Self. The Self and consciousness, I think, are names we give to the sensations produced by all of those electrical and chemical flows.

If someone began disassembling my brain one neuron at a time, depending on where the process began I might first lose a few motor skills, then some memories, then perhaps the ability to find particular words to make sentences, the ability to recognize faces, the ability to know where I was. During this slow taking apart of my brain, I would become more and more disoriented. Everything I associate with my ego and Self would gradually dissolve away into a bog of confusion and minimal existence. The doctors in their blue and green scrub suits could drop the removed neurons, one by one, into a metal bowl. Each a tiny gray gelatinous blob. Stringy with the axons and dendrites. Soft, so you would not hear the little thuds as each plopped in the bowl.

Likewise, the same doctors in their blue and green scrub suits could create consciousness by building a brain from scratch, one neuron at a time, delicately arranging the connections between neurons. The doctors might connect some of the neurons to a device that monitored their combined electrical activity. Neuron by neuron, connection by connection. At first, there would be simply noise. But at some point, presumably, a change would occur, a coherent signal, perhaps Desimone's synchronous hum, that would translate, roughly, into "Yikes, something is messing with me."

If we conceive of death as nothingness, we cannot imagine it. But if we conceive of death as the complete loss of consciousness, a view supported by the understanding of the body as an arrangement of material atoms, then we approach death in gradual stages as consciousness fades and dissolves. The distinction between life and death would no longer be an all-or-nothing proposition.

The neuroscientist Antonio Damasio has defined different levels of consciousness. The lowest level, which he calls the "protoself," is related to an organism's ability to carry out the most basic processes of life, but nothing else. An amoeba has a protoself. I would not associate this level of existence with consciousness. Almost certainly, thought and self-awareness require a minimum number of neurons, well beyond the stuff of an amoeba. Next comes "core consciousness." It is self-awareness and the ability to think and reason in the present moment, but without memories extending earlier than a few minutes into the past. Such an organism, far above an amoeba, might be able to have an understanding of the world around it and its place in that world, but it would exist only in the present. People with certain brain disorders have only core consciousness. They cannot form new memories that last more than a few minutes. They cannot remember what happened to them in the past, except for isolated periods. For the most part, they cannot recall past personal relationships or the people they loved and who loved them. They cannot make plans for the future. They are trapped in the moment.

The highest level of consciousness is "extended consciousness," which all healthy human beings possess. Here, we can remember most of our past life as well as function completely in the present. We can remember our view of the world based on past experiences, we can remember our value system as grounded in those experiences, we can remember what we like and don't like, places we've been and people we've met. Self-identity, as most psychologists understand it, probably requires extended consciousness, that is, long-term memory. These are complex issues, not fully understood.

The slow dismantling of a human brain, whether by my imaginary doctors in their scrub suits or by the deterioration of the brain in neural disease, might proceed from extended consciousness to core consciousness to the protoself. Or perhaps it might proceed in a less orderly manner, by removing chunks of extended consciousness and core consciousness here and there until nothing is left but the protoself. However it proceeds, one begins with full consciousness and ends with an amoeba-like existence, alive only by the biologists' formal definition of the word. One begins with a full life and ends with death, or the equivalent of death. And this process can happen gradually, so that there may be some awareness of the increasing loss of awareness.

Personal accounts of early dementia provide the best knowledge we have of approaching death in this manner. In the early stages of dementia, enough of the mind remains to understand and articulate what is happening. In later stages, the reporter has slipped and disappeared into the abyss of confusion. Somewhere in that intermediate nether zone, the sense of self dissolves and is gone. It's a grim subject.

Some of my own loved ones have gone through various forms of dementia. Many of us will not suffer this depressing approach to death, and I prefer not to think of it myself. But consciousness and its loss are part of my musings today on the boundary between life and death. For a person who does not believe in the afterlife, consciousness is the subject of interest. For a materialist, death is the name that we give to a collection of atoms that once had the special arrangement of a functioning neuronal network and now no longer does so.

From a scientific point of view, I cannot believe anything

other than what I have laid out above. But I am not satisfied with that picture, just as I was not satisfied with Desimone's explanation of consciousness. In my mind, I can still see my mother dancing to the bossa nova as she often did, giving her hips a jaunty shake with the beat. I still can hear my father tell his Cooshmaker joke—"It went coosh, and I can have another one up in fifteen minutes." I often wonder: Where are they now, my deceased mother and father? I know the materialist explanation, but that does nothing to relieve my longing for them, or the impossible truth that they do not exist.

I have a confession to make. Despite my belief that I am only a collection of atoms, that my awareness is passing away neuron by neuron, I am content with the illusion of consciousness. I'll take it. And I find a pleasure in knowing that a hundred years from now, even a thousand years from now, some of my atoms will remain in this place where I now lie in my hammock. Those atoms will not know where they came from, but they will have been mine. Some of them will once have been part of the memory of my mother dancing the bossa nova. Some will once have been part of the memory of the vinegary smell of my first apartment. Some will once have been part of my hand. If I could label each of my atoms at this moment, imprint each with my Social Security number, someone could follow them for the next thousand years as they floated in air, mixed with the soil, became parts of particular plants and trees, dissolved in the ocean and then floated again to the air. Some will undoubtedly become parts of other people, particular people. Some will become parts of other lives, other memories. That might be a kind of immortality.

The Ghost House of My Childhood

The distant Earth bristles with tiny imitations of houses and roads as I slide through the air in the silver ghost, a miracle of science. Now that my second parent has passed away, all things seem strange. Am I awake or asleep? I am flying back to Memphis, the place of my childhood, to settle the last of my father's affairs and to see one final time the house where we all lived.

I sit at a table at Panera Bread. After lunch, I will get into my rented car and drive out to West Cherry Circle. I was hoping my brothers would join me on this trip to our family home, but they don't want to see it again. We sold the house months ago. I look out a window toward Poplar Avenue and remember the diner that once stood across the street, the Ohman House, where my friends and I used to go late at night, after high school dances and parties, to eat hamburgers smothered with onions, hash brown potatoes, and black bottom pie.

The Ghost House of My Childhood

It's time. I get into the rented car. When I last visited the house, two years ago, my father was waiting to greet me. He sat in the den in his wheelchair, wearing a warm sweater even in April and soft bedroom slippers, an open book on his lap.

I turn onto West Cherry Circle, drive past familiar houses. Flowers are blooming, it's spring. But something is wrong. The house isn't here. There's a hole in space where the house used to be. Slowly, I inch up the driveway and park the car. Something is terribly wrong. I feel as if I'm not in my body any longer. My body is a distant, cold moon. There was a two-story house here, with pink brick walls and a porch with white posts and dormer windows. I can see right through the empty air to bushes and trees on the other side. And on the ground where the house was, new grass. Not a single brick or splinter or piece of debris.

Slowly, I get out of my car, a knot forming in my gut, somebody's gut, and I walk around the patch of grass where the house used to be. The space is too small. I stare at the driveway, follow it with my eyes as it winds down to the street, curves by the towering magnolia around which my brothers and I once chased one another with a gushing garden hose. I stare at the neighboring houses, the fence at the back of the lot, thinking that somehow I've made a mistake.

I take a step back, blink. But there is only the silent, dead air. There was a house here. There was a cosmology of lives lived here, meals of fried chicken and mashed potatoes at the wood table in the kitchen, closets of clothes, drawers, homework by the light of the maroon double lamp, cops and robbers games with my brothers, my father shaving in the morning, evenings watching TV. I try to put the house back where it was, the kitchen, the bedrooms, the closets, my father practicing his guitar, my mother dressing in front of her long mirror. I try to will it into solidity. It was here.

Some careless god has cut the ribbon of my life. The sixty-five years of the past, and the remaining years of my future. The piece that was the past has slipped away into black eternity, or perhaps into nothingness. Until this moment, I was sure that the past was still present, caught in the spaces between things, in photos, in books, in places my body had been. I try to spool back time in my mind. I walk to a spot near a disheveled azalea. Here, in this empty corner of air, I remember waking up with a bad dream and getting into my brother's bed beside him. Our beds were six feet apart, a desk against the wall, a closet, a white woolly rug on the floor. Here, where I am standing at this moment. And over there, I remember helping my father get the boat paddles out for a trip to the lake. Second floor. A closet with a dangling bulb for light. And there, the mahogany secretary with the leather-bound books, where my mother wrote letters in her back-slanted script. I can see her sitting there at the desk in her bathrobe, twitching her legs nervously under her chair.

I am trying to remember where I came from this morning—another city, another house, my wife, what precisely she said as I packed my small bag. I try to picture her face, her hair, what clothes she was wearing. I try to remember what we ate for dinner last night. Partial images slide through my mind, a scattering of words spoken. Neurobiologists say that memory isn't the replay of a video camera, but instead a pastiche of neuronal fragments gathered from here and there, wandering smells, oddly cut visual scraps, translucent experiences laid on top of one another. It's all in the electrical currents and flow of particular molecules. Neurobiologists say that connections between the billions of neurons in a human brain change over time. If so, the universe shifts and shifts and shifts in our minds.

I am remembering wrong. I wish that my brothers were here. I want to see the people who lived in my past, the piece of the ribbon that has slipped away. We could compare testimonies. They lived in this house. But their heads are not mine. They have their own billions of neurons with shifting connections. Some philosophers claim that we know nothing of the external world outside our minds—nothing compared to what sways in our minds, in the long, twisting corridors of memory, the vast mental rooms with half-open doors, the ghosts chattering beneath the chandeliers of imagination.

If there's a mismatch between what you remember and what you see ten feet away, which one is real? Chairs. Smells. Brothers. What do you know? How do you prove that the drawer that you opened this morning is the same drawer you closed at night? And the billions of neurons go spinning their tales.

I remember a moment from when I was twelve years old, as I watched my father get measured for a shirt. His tailor had come out to the house and met him in the middle bedroom downstairs—about ten feet from where I stand now. My father would have been forty-one years old, a slight man with handsome, delicate features. The tailor wrapped his tape measure around my father's neck and they talked casually as if they were old friends, laughing together. I strained to hear the exact words. I had never seen this tailor before, but he and my father were on such easy terms with each other that it brought a calm over me. A world where a friendly tailor comes out to the house to measure my father for a new shirt is a safe world. Is the world still safe? I am standing there now. I am standing here now. I am waiting, and listening.

I am imagining it all. Perhaps even myself. Or rather, the sense that I am a self, something other than the massing of atoms and molecules, the tinglings of neurons. From all of those chemical and electrical tremblings, the illusion of consciousness. "Dream delivers us to dream, and there is no end to illusion," wrote Emerson. For the moment, my body has left me. Physicists say that time is relative. Here, in this space where there once was a house, time has also dissolved. I've been tricked and defeated by time.

A landscaping truck is coming up the driveway. It parks. Two men get out of the truck with shovels, plants, bags of manure. One of them gives me a puzzled look, as if to ask why I'm here, then ignores me and sets about spreading the fertilizer and digging into the soil. Perhaps I am not here. I look at the men and imagine that I can see right through them as I see through the slab of air where the house used to be.

These guys have no conception of what was once here. They go about their business digging and placing the plants, and all they see is an empty lot. Their neurons are different from mine. They have their own cabinets of memories. Perhaps at this moment they are thinking of their own gardens and yards, places they've been, girlfriends and wives. I wonder if I'll remember these two men tomorrow. They are brief, here at this moment. For a couple of days, I might remember them as they seem now, wearing jeans and boots and dark glasses, gloves, one smoking a cigarette. The picture will grow dimmer and dimmer, until it is gone, lost like the house that was here—part of the past that does not exist.

I am back in the restaurant of a few hours ago. It is all here as I remember it. The people typing at their laptops. The blue flames in the gas fireplace. A piece of paper in my pocket says that I am flying away tomorrow. Someone I used to know sits at a table. I think it is him. "David," I say. Perhaps he does not hear me.

In Defense of Disorder

The Buddhist monks from Namgyal monastery in India engage in a ritual that involves the creation of intricate patterns of colored sand, known as mandalas. As large as three meters across, each mandala requires a couple of weeks of painstaking work, in which several monks in orange robes bend over a flat surface and scratch metallic vials. The vials extrude sand from tiny spouts, a few grains at a time, onto areas bounded by carefully measured chalk marks. Slowly, slowly, the ancient pattern is made. After the thing is completed, the monks say a prayer, pause a moment, and then sweep it all up in five minutes.

Although I haven't witnessed this particular ritual, I've seen a number of mandalas during my travels in Southeast Asia. For Buddhists, the creation and destruction of a mandala symbolizes the impermanence of Earthly existence. But the ritual also reminds me of the profound symbiosis of order and disorder at the core of our world.

Somewhat surprisingly, nature not only requires disorder but thrives on it. Planets, stars, life, even the direction of time all depend on disorder. And we human beings as well. Especially if, along with disorder, we group together such concepts as randomness, novelty, spontaneity, free will, and unpredictability. We might put all of these ideas in the same psychic basket. Within the oppositional category of order, we can gather together notions such as systems, law, reason, rationality, pattern, predictability. While the different clusters of concepts are not mirror images of one another, like twilight and dawn, they have much in common.

Our primeval attraction to both order and disorder shows up in modern aesthetics. We like symmetry and pattern, but we also relish a bit of asymmetry. The British art historian Ernst Gombrich believed that although human beings have a deep psychological attraction to order, perfect order in art is uninteresting. "However we analyze the difference between the regular and the irregular," he wrote in *The Sense of Order* (1979), "we must ultimately be able to account for the most basic fact of aesthetic experience, the fact that delight lies somewhere between boredom and confusion." Too much order, we lose interest. Too much disorder, and there's nothing to be interested in. It's something about the human mind. My wife, a painter, always puts a splash of color in the corner of her canvas, off balance, to make the painting more appealing. Evidently, our visual sweet spot lies somewhere between boredom and confusion, predictability and newness.

Human beings have a conflicted relationship to this order-disorder nexus. We are alternately attracted from one to the other. We admire principles and laws and order. We embrace reasons and causes. We seek predictability. Some of the time. On other occasions, we value spontaneity, unpredictability, novelty, unconstrained personal freedom. We love the structure of Western classical music, as well as the freewheeling runs or improvised rhythms of jazz. We are drawn to the symmetry of a snowflake, but we also revel in the amorphous shape of a high-riding cloud. We appreciate the regular features of purebred animals, while we're also fascinated by hybrids and mongrels. We might respect those who manage to live sensibly and lead upright lives. But we also esteem the mavericks who break the mold, and we celebrate the wild, the unbridled, and the unpredictable in ourselves. We are a strange and contradictory animal, we human beings. And we inhabit a cosmos equally strange.

You can see the creative tension of the order-disorder nexus in our science versus our art. In his law of floating bodies, formulated in 250 BC, Archimedes prefigured the coming age of science when he expressed one of the first quantitative laws of nature: "Any body wholly or partially immersed in a fluid experiences an upward force equal to the weight of the fluid displaced." In other words, a body sinks just to the level where the weight of the displaced fluid equals the weight of the body. To verify this elegant law, Archimedes would have done the experiment over and over with various objects of different shapes and sizes, and with different liquids such as water and mercury. (Scales were available in the Greek *agora* for weighing wheat, salted fish, glass, copper, and silver.)

Evidently, the world of masses and forces was logical, rational, quantifiable, predictable. Yet two centuries earlier, Socrates—that wandering sage whom Plato and others described as resembling a satyr more than a man, short and stocky, with a pug nose and bulging eyes—celebrates the creative power of madness: "He who, having no touch of the Muses' madness in his soul, comes to the door and thinks that he will get into the temple by the help of art—he, I say, and his poetry are not admitted; the sane man disappears and is nowhere when he enters into rivalry with the madman." Creativity has always been associated with novelty, surprise, and what psychologists and neuroscientists call *divergent thinking:* the ability to explore many different avenues and solutions to a problem in a spontaneous and non orderly fashion. *Convergent thinking,* by contrast, is the more logical and orderly step-by-step approach to a problem. The French mathematician Henri Poincaré in 1910 described the gestation of one of his mathematical discoveries as a dance between the two:

> For 15 days, I strove to prove that there could not be any [mathematical] functions like those I have since called Fuchsian functions. I was then very ignorant; every day I seated myself at my work table, stayed an hour or two, tried a great number of combinations and reached no results. One evening, contrary to my custom, I drank black coffee and could not sleep. Ideas rose in crowds; I felt them collide until pairs interlocked, so to speak, making a stable combination. By the next morning...

Undoubtedly, some of our creativity is ignited by a synthesis of convergence and divergence, working together in symphony.

The critical role of disorder in nature was not articulated until two thousand years after Socrates praised the mad poet. The task fell to the German physicist Rudolf Clausius. He was born in 1822 in Pomerania, a region split between Germany and Poland, and educated at the University of Berlin. Perhaps under the influence of his religious father, a clergyman, Clausius led a principled life. "A chief characteristic was his sincerity and fidelity" was how his brother, Robert, described Clausius at his death in 1888. "Every kind of exaggeration was opposed to his nature."

Like Einstein, Clausius was a theoretical physicist—that is, all of his work, including his seminal work on disorder, consisted of mathematical feats performed with pencil and paper. Clausius's great paper on disorder, "On the Moving Force of Heat" (1850), was published the same year that he became a professor of physics at the Royal Artillery and Engineering School in Berlin. In that paper, Clausius showed that change in the physical world is associated with the inevitable movement of order to disorder. Indeed, without the potential of disorder, nothing in the cosmos would ever change—like a row of upright dominoes held rigidly in place, or like a completed Buddhist mandala locked in a bank vault, safe from the brooms of the monks of Namgyal. "Heat" occurs in the title of Clausius's paper because increasing disorder is often associated with the transfer of heat from hot bodies to cold—but the concept is more general. In a later paper, Clausius coined the term *entropy* as a quantitative measure of disorder. The word comes from the Greek έν (en), meaning "in", and τροπή (tropē), meaning "transformation." It is the increase of entropy that is linked to transformation, movement, change in the world. The more disorder, the more entropy. The last two sentences of Clausius's 1850 paper:

1. The energy of the Universe is constant.
2. The entropy of the Universe tends toward a maximum.

Order inevitably yields to disorder, and entropy increases until it cannot increase any further. It is this movement that drives the world. Clean rooms become dusty. Temples slowly crumble. As we grow older, bones grow brittle. Stars eventually burn out, emptying their hot energy into the coldness of space—but while doing so, they provide warmth and life to surrounding planets. We live off this relentless increase of disorder.

Even something as fundamental as the direction of time is governed by the movement of order to disorder, as I discussed in "What Came Before the Big Bang?" Because everything passes from order to disorder as we march toward the future. One might say that the forward direction of time *is* the increase in disorder. Indeed, without these changes, we'd have no way of telling one instant from the next. There would be no clocks, no flights of birds, no leaves slipping through the air as they dropped from trees, no breathing out and in. The universe would be a still photo for all of eternity.

Disorder is also the answer to the profound question Why is there *something* rather than *nothing?* (Such questions keep physicists and philosophers up at night.) Why does material of any kind exist, rather than pure energy? From a scientific perspective, the question relates to the existence of antiparticles, predicted in 1931 and then discovered in 1932. Every subatomic particle, such as the electron, has an antiparticle twin—identical to the first, except with opposite electrical charge and certain other qualities. Which of the pair we call the "particle" and which we call the "antiparticle" is a matter of convention, like the North and South Poles. When they meet, particles and their antiparticles annihilate each other, leaving nothing but pure energy.

If there were an equal number of particles and their antiparticles in the infant universe, as one would expect in a completely symmetrical universe, all matter would have been obliterated billions of years ago, leaving nothing but pure energy. No stars, no planets, no people—or any other solid material. So why are we here? Why haven't all the particles disappeared along with their antiparticle partners?

The answer to this physicists' conundrum came in 1964. In very delicate experiments, we discovered that particles and antiparticles do not behave in *exactly* the same way. Rather, there is a slight asymmetry in how they interact with other particles, so that immediately after the creation of the universe, particles and their antiparticles were not produced and destroyed in equal numbers. After the mass annihilations of particles with their antiparticle partners, some particles would remain, like a surplus of boys sitting on the bench at a school dance. Those remaining particles and the asymmetry that produced them is why we exist.

Disorder isn't present only in the minutiae of how matter organizes itself. It also runs deep within the structures of life itself. Perhaps the most well known example of disorder in biology is the shuffling of genes—both by mutation and by the transfer of genes from viruses and other organisms. Through these random processes, living organisms try out different bodily architectures that might have never been sampled otherwise. These spins of the genetic roulette wheel aren't planned, and their outcomes can't be known in advance. But without them, biology would be stuck with a small number of inflexible designs. Many organisms would die out, unable to adapt to changing environmental conditions.

Another significant way that disorder makes itself known in biology occurs via a process called *diffusion*. Here, a lumpy gob of matter or energy is automatically smoothed out by the random collisions of atoms and molecules. You can see this for yourself if you pour a bucket of hot water into a cool bath. At first, the bath will have a hot region surrounded by a cool region. But the hot water will quickly mix with the cool until the bath comes to a uniform temperature. That's diffusion. To paraphrase Clausius, diffusion doesn't cost any energy, but it increases disorder—in this case, mixing heat—which drives transformation and change. Without random molecular collisions, diffusion would not occur. The hot water would remain at one side of the tub and the cool water on the other.

Diffusion is a key mechanism for transporting vital substances throughout the body. Take oxygen, the essential gas for energy production. With each inhalation, we produce a high concentration of oxygen in our lungs. The tiny blood vessels embedded in the lungs have a relatively low amount of oxygen. That allows the vital gas to "diffuse" from the lungs to the blood, and then, for the same reason, from the blood to individual cells throughout the body. Such directed movement results from random collisions, tending to transport oxygen molecules from areas of high to low oxygen concentration. Without random bumps and knocks, oxygen in the lungs would remain trapped in the lungs, and the cells of the body would suffocate.

But none of these examples from the microscopic domain, including Clausius's deep pronouncements on entropy, explain our psychological attraction to both order and disorder—our honoring of both the respectable and the mavericks among us. There seems to be something deep in our psyche, something primeval, imprinted in us eons before Clausius or Socrates. Perhaps the embrace of these opposites conferred an adaptive advantage on our ancestors, many millions of years in our past. From an evolutionary point of view, order implies predictability, patterns, repeatability—all of which allow us to make good predictions. And predictions are useful for knowing when game will run through the forest, or when crops should be planted. The benefit to our survival is obvious. More unexpected, perhaps, is how attentiveness to surprise, chance, and novelty can also confer an advantage. If we get too complacent with our routine, we can't react when things change, when the tiger suddenly appears on the path that we have walked a thousand times without mishap. And we would not take risks, for fear of departing from our familiar routines. So it makes sense that we've developed a desire for both the predictable and the unpredictable.

If an appetite for novelty conferred a survival benefit on our ancestors, perhaps it should show up in our genes. Researchers have recently discovered a variation (allele) of a gene called DRD4-7R—or more arrestingly, "the wanderlust gene." It occurs in about 20 percent of the population and appears to be associated with a penchant for exploration and risk. It makes sense that we'd want most members of our tribe to stay at home, follow routine, tend to the hearth. Yet we also need a few others to venture forth on risky expeditions in search of new hunting grounds and unexpected opportunities. "We have evidence to suggest that the same allele involved in the personality trait of novelty-seeking and impulsivity was also involved in being pro-risk in financial situations," says Richard Paul Ebstein, professor of psychology at the National University of Singapore and one of the leading researchers of DRD4-7R. "People who have that allele appear to be more risk-prone." Other biologists rightly point out that it's unlikely that any single gene could control a trait such as risk taking and novelty seeking—but a group of genes working together might just do so.

Since both order and disorder evidently benefit human beings, it's worth re-examining our inclination, at least in the West, to divide everything into polar opposites, with an assumed hierarchy of value and unstated preferences—productivity versus laziness, rationality versus irrationality, hot and cold, smooth and rough, white and black. Perhaps, instead, we should view such opposites in terms of a useful balance.

The Chinese have long understood this idea in terms of the ancient Confucian concept of yin and yang: all things exist as inseparable and contradictory opposites. Yin is associated with the feminine, dark, north, old, soft, cold, while yang with masculine, light, south, young, hard, warm. The symbol of yin and yang—two entangled swirls, one black and one white, equal in size, each with a dot of the other color within it—suggests that the two exist in harmony, with neither dominant over the other. Meanwhile, Western thought typically attempts to simplify this baffling world by dividing everything into two. That works for a while, until we look more closely and discover the real complexity lurking underneath. If eventually we are able to stand on higher ground, we once again find simplicity and harmony. The cosmos sings order, and it also sings disorder. We human beings seek predictability, and we also yearn for the new. Embrace these necessary contradictions, say the Confucians. Perhaps Pascal's Nothingness and Infinity are also part of the yin/yang balance.

It is the end of my day, and I am listening to Anton Bruckner's Ninth Symphony, which the Austrian composer began writing in 1887. The symphony opens with a continuous unfolding of the themes. The second movement, the Scherzo, feels sinister, as if some dark secret is being withheld. But I find myself mesmerized by a section of the third movement, the Adagio. After a haunting and harmonious melody from the strings (perhaps promising to reveal the dark secret), the sounds become increasingly discordant, building in volume, until we hear a thunderclap of the horns, jagged and dissonant, followed by more clashes, like tidal waves pounding the shore. Then a moment of silence. The strings pick up again, quiet and lyrical. This alternation of the melodious with the dissonant continues until the end of the movement. And I wonder if the harmonious sections of the piece would be quite so beautiful if not juxtaposed with the unharmonious, the light with the dark, the smooth with the rough. The orderly with the seemingly disorderly. And of course Bruckner himself— a chance event like all of us, a random collision of cells bringing forth improbable life in this improbable universe.

Miracles

Then Moses stretched out his hand over the sea; and the Lord caused the sea to go back by a strong east wind all that night, and made the sea into dry land, and the waters were divided. So the children of Israel went into the midst of the sea on the dry ground, and the waters were a wall to them on their right hand and on their left."

These words from Exodus describe one of the most famous miracles in the Bible. Never before and never since, in any sea or ocean on Earth, have winds created a passageway through which people could walk. In scientific terms, such an event would require a sustained and *highly directed* column of wind blowing at hurricane force, a phenomenon that could be created only on a small scale in a human-made wind tunnel of the twentieth century. But the parting of the Red Sea occurred three thousand years ago. It was ordered up by Moses and delivered by God. It was a "miracle," at odds with the behavior of nature, beyond nature, a "supernatural" event, inexplicable except by recourse to divine intervention.

A 2013 Harris poll found that 74 percent of Americans surveyed believe in God, and 72 percent believe in miracles. Miracles are usually associated with the actions of gods or other divine beings, and they occur not only in Judaism and Christianity but in all of the major religions of the world. In Islam, Mohammad split the Moon. In Hinduism, when Saint Jnanadeva was told that he was not qualified to recite the Vedas, he placed his hand on a water buffalo, which proceeded to chant Vedic verses. Most Buddhists believe that all living creatures experience a cycle of deaths and rebirths, appearing in new bodies and passing through various nonphysical realms on the way.

Miracles, by definition, lie outside science. Miracles are incompatible with a rational picture of the physical world. Nevertheless, even in our highly scientific and technological society, with most of us profiting enormously from cell phones and automobiles and other products of science, a large fraction of the public believes in miracles. Most of us do not ponder that contradiction. One of my aunts was certain that her dead father visited her house and spoke to her every few months, and she got a tape recorder—a device of science—to document his voice. (Whereupon the ghostly visits ceased.)

Miracles come from the world of imagination, of dreams, of desire; science from the world of practicality, of logic, of orderly control. I've always been fascinated by our ability to live simultaneously in both of these apparently opposing worlds. Evidently, each reflects something deep and essential inside of us.

While miracles that seemingly defy nature have long featured in human history, so too has our quest to codify nature, to enshrine its properties into so-called laws of nature. The laws of nature are usually stated in mathematical form. The premier example in the history of science is Isaac Newton's law of gravity: the strength of the gravitational force between two masses doubles when either mass doubles and increases fourfold when the distance between them is halved. (In mathematical form, $F = Gm_1 m_2 /d^2$.) It is a rule that Newton derived to explain the orbits of the planets, and it can be used to predict how masses will affect one another through their mutual gravity anywhere in the universe. As an application of Newton's law: since the Moon is ¼ the the size of the Earth and roughly 1/100 the mass, you would weigh about ⅙ as much on the Moon as you do on Earth. (This little-known fact does not appear in the diet books I've seen.)

As another example, which you can test for yourself: drop a weight to the floor from a height of 4 feet and time the duration of its fall. You should get about 0.5 seconds. From a height of 8 feet, you should get about 0.7 seconds. From a height of 16 feet, about 1 second. Repeat from several more heights and you will discover that the time exactly doubles with every quadrupling of the height, a rule found by Galileo in the seventeenth century. (Mathematically, $t = $ constant $\times \sqrt{h}$.) With this rule, you can now predict the time to fall from any height. You have witnessed, firsthand, the regularity of nature.

Why should nature be lawful? One can imagine a universe in which events happened at random, without any justification or regularity. A wheelbarrow might suddenly float in the air. Day might turn into night and back to day at arbitrary moments. In such a universe, of course, scientists would be out of business. Not only do scientists depend on the regularity and logic of nature, most scientists would argue that an irrational and unmathematical universe could not exist. Undoubtedly, the lawfulness of nature, and especially our ability to find those laws—from Archimedes to Newton to Einstein—has brought us human beings a sense of power, a sense of comfort and security, and a sense of control.

Beyond the personal wishes of scientists, the concept of a lawful nature has proven enormously useful. The regular and predictable cycles of the seasons allowed for the development of agriculture. The consistent properties of materials allowed for the development of industry. The repeatable production of T-lymphocytes and other antibodies when exposed to the vaccinia virus allowed the eradication of smallpox, one of the greatest killers of human beings throughout history.

In addition to these practical applications, science has also been able to explain and predict the more esoteric behavior of nature to high accuracy. For example, the orbit of Mercury rotates a slight amount more than could be accounted for by Newton's seventeenth-century law of gravity. The tiny discrepancy, 0.012 degrees per century, was successfully calculated by Einstein's modern theory of gravity, general relativity.

Finally, scientists—and to a great extent the population at large—now believe that these laws are discoverable by human beings. That belief was not always so. For centuries, people thought that various forms of knowledge, including knowledge about the workings of nature, were the sole province of God, off-limits to human understanding. The great success of modern science has challenged that view, whether or not one believes in God. In a sense, the success of science, our own human enterprise, has empowered us to proclaim that nature is lawful.

All of the above progress has led to what one might call the Central Doctrine of Science: All properties and events in the physical universe are governed by laws, and those laws hold true at every time and every place in the physical universe. Scientists do not explicitly discuss this doctrine. It is simply assumed. When I was a graduate student in physics, my thesis advisor never mentioned the doctrine, but it was implicit in everything he did in his own professional work and in the guidance he gave to his students. One of my first research problems as a physicist concerned the behavior of very hot gas at the centers of galaxies. For a sufficiently hot gas, electrons and their antiparticles can be created out of the immense thermal energy. Early on, I had to write down the equations governing how matter can be created from energy, a result expressed by Einstein's famous $E = mc^2$ and confirmed in numerous laboratories on Earth. At no point of my calculations did I have any doubt that the same equations applied to distant galaxies, millions of light-years away.

Philosophers debate about whether the "laws of nature" are mere *descriptions* of nature or *necessities* of nature, the latter being rules that nature must obey without exception. The Central Doctrine of Science, and the view of most scientists, is that the laws are necessities.

Throughout time, human beings have had a complex and evolving conception of nature. Mother Nature exists in every culture on Earth. She was known as Gaia in ancient Greece, Terra Mater in ancient Rome. Ninsun in ancient Mesopotamia. In India, Gayatri. In Thailand, Phra Mae Thorani. The Māori call her Papatuanuku. In early times, when Nature was personified, she could be angry and vengeful, loving, indifferent. Many religious traditions today continue to associate various deities with nature. The 330 million gods in Hinduism permeate nature. The gods, of course, are not bound by the rules found by science, or any other rules. In such a world view, the boundaries are blurred between the rational and the irrational, the predictable and the unpredictable, the ordinary and the miraculous.

But even in the Judaic-Christian beliefs and traditions, those boundaries are blurred. How else to explain that more than two-thirds of the American public believes in miracles while, at the same time, trusts in science every time they turn the wheel of their car to make a slight correction in direction while traveling at sixty miles per hour on the highway? Having myself lived both in the territory of science (as a physicist) and in the territory of the arts (as a novelist), I would like to offer an opinion about how and why such apparent contradictions and blurrings occur.

The miraculous has meaning only by contrast to the non-miraculous, the ordinary, the normal behavior of nature. In our modern world—with climate-controlled buildings, asphalt highways, artificial turf, computers and iPhones with which we can talk to images of our friends thousands of miles away—it seems that most of us have only a vague idea of what is "natural" and what is "unnatural." Rarely do we observe the natural world without some mediation by an artificial device. Even in science, astronomers no longer look directly through the eyepiece of a telescope, but instead see images collected by digital devices called CCDs and presented on computer screens.

In recent years, the environmental movement has somewhat increased our awareness of nature. In his book *Earth in the Balance,* Al Gore writes, "The disharmony in our relationship to the Earth, which stems in part from our addiction to a pattern of consuming ever-larger quantities of the resources of the Earth, is now manifest in successive crises, each marking a more destructive clash between our civilization and the natural world." But even environmental consciousness has not much changed the disconnection between human beings and nature. Most of us still live in cities where we cannot see the star-spangled dark of the night sky. Most of us heat our houses in winter and cool them in summer, insulating ourselves from the cycles of the seasons.

But there is to me a far more compelling explanation for our ability to hold both the miraculous and the nonmiraculous in our heads at the same time. Many of us, consciously or unconsciously, believe in some kind of a spiritual universe existing alongside the physical universe. Miracles, then, involve an interaction of those two distinct forms of existence.

Let me mention two exceptions to this division. In pantheism, a philosophy popularized by Baruch Spinoza in the seventeenth century, there's no separation between the physical and spiritual universes. There is only a single universe. Nature brims with God. Nature has no boundaries. Nature is everything. In such a situation, the so-called scientific laws of nature describe only one aspect of nature. In the other aspect, the divine, events occur that are indescribable and unpredictable by science. Another exception is deism, in which the two universes are distinct, but God does not act in the physical universe. There is no intersection. God set the universe in motion and then sat down. Thus, in deism, miracles cannot occur. Deism, which gained prominence in the Enlightenment, provided a way for people to reconcile their religious beliefs with the rise of modern science.

The more challenging world view is that in which the spiritual and physical universes are distinct but engage with each other from time to time, in the form of miracles, which break the boundaries of an otherwise law-based existence. In this view, beings and events in the spiritual universe sometimes cross over and appear in the physical universe. Prime examples include the parting of the Red Sea, the Resurrection of Christ, and the splitting of the Moon by Mohammad. At a more mundane level, many of us report experiencing "little" miracles in our day-to-day lives, such as memories of existence in a previous life, or premonitions of future events that then happen, or socks that have disappeared into hyperspace.

Even some scientists believe in such crossovers. Owen Gingerich, professor emeritus of astronomy and of the history of science at Harvard University, said to me: "I believe that our physical universe is somehow wrapped within a broader and deeper spiritual universe, in which miracles can occur. We would not be able to plan ahead or make decisions without a world that is largely law-like. The scientific picture of the world is an important one. But it does not apply to all events." I would estimate that something like 3 to 5 percent of all scientists share Gingerich's view. Such scientists—who are clearly in the minority—believe that science and the lawfulness of nature hold true *most* of the time, but occasionally God intervenes in the physical world and acts in a way that cannot be analyzed by science.

Belief in a spiritual universe, I would suggest, arises to a large extent from a human desire for *meaning*, meaning both in our individual lives and in the cosmos as a whole. While science provides the psychological comfort of order, rationality, and control, it does not provide meaning. Such deep philosophical questions as "Why am I here?" "What is the purpose of my life?" "What is the meaning of this strange cosmos I find myself in?" and such moral questions as "Is it right to kill an enemy soldier in time of war?" and "Is it right to steal in order to feed my family?" cannot be answered by science. Yet these questions are vital to our mental and emotional lives. The spiritual universe is the place we turn to for answers to these questions, the realm that contains eternal truths and guidance, the realm that has some kind of permanent existence, in contrast to the fleeting moment of our mortal lives. In such a realm, logic, rationality, and regularity are not even part of the vocabulary.

A spiritual universe may or may not include God. However, it is usually associated with religion. In his landmark book *The Varieties of Religious Experience: A Study in Human Nature* (1902), the Harvard philosopher and psychologist William James characterized religion, in its broadest terms, as "the belief that there is an unseen order, and that our supreme good lies in harmoniously adjusting ourselves thereto." The "order" in James's conception of religion helps provide meaning. The order must be unseen, since much of what we see in the world of human affairs is chaotic and hard to understand with rational thought. The hypothesized order provides comfort and security. The hypothesized order suggests that there is some purpose in the cosmos. The order comes from outside the physical universe. It comes from the spiritual universe. Paradoxically, for believers, that same unseen order sometimes chooses to violate the orderly laws of nature and produce a miracle.

I confess that I myself do not believe in miracles. I have sometimes wondered why this disbelief has always been strong, even from a young age. I suppose my views have been shaped in part by demonstrating to my own satisfaction that the physical world is a lawful place. I remember that as a boy of twelve or thirteen, among my many scientific projects, I began making pendulums by tying a fishing weight to the end of a string. I built pendulums of different lengths and timed their swings with a stopwatch. I had read in *Popular Science* or some book that the period of the pendulum—the time it takes to make one complete swing—was proportional to the square root of the length of the string. I personally verified that formula and then used it to predict the periods of new pendulums even before I had made them. How amazing, I thought, that this simple formula worked over and over again—at my house, at my friends' houses, anywhere at all. By contrast to the erratic ups and downs and unpredictable behavior of my brothers and mother and father, nature was reliable.

As I got older, I learned more and more about the laws of nature, about what scientists did, and I didn't see any evidence of a supernatural world. It seemed to me, and still seems to me, that everything we experience in the physical world can be explained in terms of repeatable and universal laws of nature. We certainly don't have a complete set of the laws of nature, but I join most scientists in believing that a complete set of laws exists. The history of science has been a history of constant progress in discovering the laws of nature.

At the same time, it seems extremely unlikely to me that there is another kind of reality that exists outside the physical universe but that can enter our time and space at will. I simply have seen no evidence for such a thing. We all have to come to our own view of the world, based on the experiences of our own and the people we trust. In this regard, I have always put stock in Occam's razor. Among competing hypotheses to explain events, I go with the simplest, the one that requires the least number of assumptions, until that hypothesis is proven wrong. If events in the physical universe can be explained by laws of nature, then why invoke anything beyond nature? To my mind, the parting of the Red Sea and other reported miracles are not documented and confirmed. Furthermore, they contradict the reality that I have come to embrace through countless personal experiences with nature, big and small, from my childhood experiments to my research in physics to my everyday life in the world.

All of that said, I still consider myself a spiritual person. By spirituality, I mean belief in things that are larger than myself, appreciation of beauty, commitment to certain rules of moral behavior, such as the Golden Rule. Spirituality does not require belief in miracles.

My wife and I spend summers on a small island in Maine, far from any town. At night, the skies are quite dark. Sometimes, when there is no wind blowing and the tidal flow is small and the ocean is very still, I can see the reflection of the stars in the water. At such moments, the water looks like a dark carpet with a million tiny sparkles of light, which gently bob and ripple with each passing wave. Even though I know all the science, I am totally mesmerized and awed. For me, that is miracle enough.

Our Lonely Home in Nature

The tornadoes that devastated Arkansas and several other states in 2014, turning homes into matchsticks and killing dozens of people, and the deadly mudslide in Washington State that same year, demonstrate once again the unimaginable power of nature. Of course, we have seen this before. The 2004 earthquake and tsunami in the Indian Ocean, killing over 250,000 people in Indonesia and other countries. Hurricane Katrina in 2005, ending the lives of at least 1,800 people and wreaking nearly $100 billion in property damage. The 2011 tsunami in Japan, drowning over 18,000 souls. And, of course, the coronavirus gripping the world at this moment in May 2020, as I now revise this essay.

After each of these disasters, we grieve over the human lives lost, the innocent people drowned or crushed or infected without warning as they slept in their beds or worked in their fields or sat at their office desks. We feel angry at the scientists and policy makers who didn't foresee the impending calamity or, if forewarned, failed to protect us. Beyond the grieving and anger is a more subtle emotion. We feel betrayed. We feel betrayed by nature. Aren't we a part of nature, born in nature, sustained by the food brought forth by nature, warmed by the natural Sun? Don't we take pleasure in walking through grassy fields and sitting barefoot on the edge of the sea? Don't we have a deep spiritual connection with the wind and the water and the land that Emerson and Wordsworth so lovingly described, that Turner and Constable painted in scenes of serenity and grandeur? How could Mother Nature do this to us, her children?

Yet despite our strongly felt kinship and oneness with nature, all the evidence suggests that nature doesn't care one whit about us. Tornadoes, hurricanes, floods, earthquakes, volcanic eruptions, pandemics happen at moments and places without the slightest consideration of human inhabitants.

I remember the first time I encountered the insensible power of nature. My wife and I had chartered a small sailboat for a two-week holiday in the Greek isles. After setting out from Piraeus, we headed south and hugged the coast, which we held three or four miles to port. With binoculars, we could just make out the glinting of houses on the land, fragments of buildings. Then we passed the tip of Cape Sounion and turned west toward Hydra. Within a couple of hours, both the land and all other boats disappeared. Looking around in a full circle, all we could see was water, extending out and out in all directions until it joined with the sky. At first, I felt elation. Then I felt fear. Because during the summer season, the Aegean Sea is plagued by a fierce, dry wind called the meltemi, which can appear without warning in clear air and be upon you in minutes with great waves and gales. At any moment, a wall of water and wind could lunge from the horizon, wash over the boat, and drown my wife and me. I realized that there was no compassionate overseer or oceanic consciousness to prevent that from happening. To the great expanse of water, my wife and I were just additional pieces of flotsam and jetsam. And I remembered someone I knew who had been walking along the shore in Alaska one day and was suddenly swept away by a crashing wave.

I believe that our comfort with nature is an illusion. Certainly, we are part of nature, but does nature care about us? Here on Earth, even with our earthquakes and storms, we have no conception of the range and the power of nature. In many other parts of the cosmos, conditions of temperature and atmosphere and gravity are far more extreme than on Earth and quite inhospitable to life. On planet Mercury, for example, the temperature is 800 degrees. On Neptune, it is −328. On Uranus, the winds exceed 500 miles per hour. There are dead stars so compact that a penny on their surface would weigh over a hundred thousand tons. In the last decade, we have discovered over a thousand planets outside our solar system, many with environments far different from Earth's. One world is apparently covered entirely with water and has an atmosphere of thick steam. Another world orbits its central star in a mere nine hours. (Its year is less than one Earth day.)

For all of recoded history, humankind has had a conflicted view of nature. In ancient times, we made awesome and frightening gods of the natural elements. The Babylonian-Assyrian god of storms, Adad, brought rain to the crops but also caused havoc and death on land and on sea. Vulcan, the god of fire, both created and destroyed and was sometimes invoked to annihilate one's enemies. In Chinese thought, especially in Taoism, we are advised to follow the flow of nature's rhythms for moral and physical health. So close to nature are we in some mythologies that human beings are regularly transformed into other animals and even inanimate matter. In Aztec mythology, the twin volcanoes Popocatépetl and Iztaccíhuatl were once human lovers, later turned into mountains by the gods. In the other direction, nature is constantly given human qualities. Wordsworth wrote that "nature never did betray the heart that loved her." Mother Nature suckles and comforts us in every culture on Earth. In the twentieth and twenty-first centuries, some environmentalists have claimed that the entire Earth is a single ecosystem, a "superorganism" called Gaia.

I would argue that we have been fooling ourselves. Nature, in fact, is *mindless*. Nature is neither friend nor foe, neither malevolent nor benevolent.

Nature is purposeless. Nature simply is. We may find nature beautiful or terrible, but those feelings are human constructions. Such utter and complete mindlessness is hard for us to accept, as creatures with minds. We feel such a strong connection to nature. But the relationship between nature and us is one-sided. There is no reciprocity. There is no mind on the other side of the wall. That absence of mind, coupled with so much power, is what so frightened me on the sailboat in Greece.

The 2014 report by the United Nations Intergovernmental Panel on Climate Change documents the damage now being done by human-created greenhouse gases and global warming, the changes in weather patterns, the rise of sea levels, the droughts and the storms and the dangers to human habitation and agriculture. In reacting to the report, we should not be concerned about protecting our planet. Nature can survive far more than what we can do to it and is totally oblivious to whether *Homo sapiens* lives or dies in the next hundred years. Our concern should be about protecting ourselves. Because we have only ourselves to protect us.

Is Life Special?

A rocket powered by kerosene and liquid oxygen and carrying a scientific observatory blasted off into space at 10:49 p.m., March 6, 2009 (by local calendars and clocks). The launch came from the third planet out from a G-type star, 25,000 light-years from the center of a galaxy called the Milky Way, itself located on the outskirts of the Virgo Cluster of galaxies. On the night of the launch, the sky was clear, with no precipitation or wind, and a temperature of 292 degrees by the absolute temperature scale. Local intelligent life-forms cheered the launch. Shortly after the blastoff, the government agency responsible for spacecraft, the National Aeronautics and Space Administration, wrote in the global network of computers: "Our team is thrilled to be a part of something so meaningful to the human race—Kepler will help us understand if our Earth is unique or if others like it are out there."

The above account might have been written by an intelligent life-form located on exactly the kind of distant planet that Kepler would search for. Named after the Renaissance astronomer Johannes Kepler, the observatory was specifically designed to search for planets outside our solar system that would be "habitable"—that is, neither so near their central star that water would be boiled off, nor so far away that water would freeze. Most biologists consider that liquid water is a precondition for life, even life very different from that on Earth. Kepler surveyed about 150,000 Sun-like stellar systems in our galaxy and discovered over a thousand alien planets. Although the satellite stopped functioning in 2013, its enormous stockpile of data is still being analyzed. For centuries, we human beings have speculated on the possible existence and prevalence of life elsewhere in the universe. For the first time in history, we can begin to answer that profound question.

At this point, the results of the Kepler mission can be extrapolated to suggest that something like 10 percent of all stars have a habitable planet in orbit. That fraction is large. With a hundred billion stars just in our galaxy alone, and so many other galaxies, it is highly probable that there are many many other solar systems with life. From this perspective, life in the cosmos is common.

However, there's another, grander perspective from which life in the cosmos is rare. That perspective considers *all* forms of matter, both animate and inanimate. Even if all "habitable" planets (as determined by Kepler) do indeed harbor life, the fraction of all material in the universe in living form is fantastically small. Assuming that the fraction of planet Earth in living form, called the biosphere, is typical of other life-sustaining planets, I estimate that the fraction of all matter in the universe in living form is roughly one-billionth of one-billionth. Here's a way to visualize such a tiny fraction. If the Gobi Desert represents all of the matter flung across the cosmos, living matter is only a few grains of sand in that desert. How should we think about this extreme rarity of life?

As I have commented in previous chapters, throughout history most of us have considered ourselves and other life-forms to contain some special, nonmaterial essence that is absent in nonliving matter and that obeys different principles. The eighth-century BC Egyptian royal official Kuttamuwa built an eight-hundred-pound monument to house his immortal soul and asked that his friends feast there after his physical demise to commemorate him in his afterlife. The eleventh-century Persian polymath Avicenna argued that since we would be able to think and to be self-aware even if we were totally disconnected from all external sensory input, there must be some nonmaterial soul inside of us. These are all "vitalist" ideas.

Modern biology has challenged the theory of vitalism. In 1828, the German chemist Friedrich Wöhler synthesized the organic substance urea from nonorganic chemicals. Urea is a byproduct of metabolism in many living organisms and, before Wöhler's work, was believed to be uniquely associated with living beings. Later in the century, the German physiologist Max Rubner showed that the energy used by human beings in movement, respiration, and other forms of activity is precisely equal to the energy content of food consumed. That is, there are no hidden and nonmaterial sources of energy that power human beings. In more recent years, the composition of proteins, hormones, brain cells, and genes has been reduced to individual atoms, without the need to invoke nonmaterial substances.

Yet, polls of the American public show that three-quarters of people believe in some form of life after death. Surely, this belief too is a version of vitalism. If our bodies and brains are nothing more than material atoms, then, as Lucretius wrote two millennia ago, when those atoms disperse as they do after death, there can be no further existence of the living being that once was.

Paradoxically, if we can give up the belief that our bodies and brains contain some transcendent, nonmaterial essence, if we can embrace the idea that we are completely material, then we arrive at a new kind of specialness—an alternative to the specialness of vitalism. We are special material. Not special because our atoms are different from atoms in rocks and water, and not special because we have a nonmaterial essence inside us, but special because our atoms are arranged in a special way as to create life, and consciousness. We humans living on our one planet wring our hands about the brevity of our lives and our mortal restraints, but we do not often think about how improbable it is to be alive at all. Of all the zillions of atoms and molecules in the universe, we have the privilege of being composed of those very, very few atoms that have joined together in the special arrangement to make living matter. We exist in that one-billionth of one-billionth. We are that one grain of sand in the desert.

The two tramps in Beckett's *Waiting for Godot,* placed on a minimalist stage without time or without space, waiting interminably for the mysterious Godot, capture our bafflement with the meaning of existence. Estragon: "What did we do yesterday?" Vladimir: "What did we do yesterday?" Estragon: "Yes." Vladimir: "Why... (Angrily) Nothing is certain when you're about." Of course, there are questions that do not have answers.

But if we can manage to get outside of our usual thinking, if we can rise to a truly mind-bending view of the cosmos, there's another way to think of existence. In our extraordinarily entitled position of being not only living matter but conscious matter, we are the cosmic "observers." We are uniquely aware of ourselves and the cosmos around us. We can watch and record. We are the only mechanism by which the universe can comment on itself. All the rest, all those other grains of sand in the desert, are dumb, lifeless matter.

Of course, the universe does not need to comment on itself. A universe with no living matter at all could function without any trouble—mindlessly following the conservation of energy and the principle of cause and effect and the other laws of physics and biology. A universe does not need minds, or any living matter at all. (Indeed, in the recent "multiverse" hypothesis endorsed by many physicists, the vast majority of universes are totally lifeless.) But in this writer's opinion, a universe without comment is a universe without meaning. What does it mean to say that a waterfall, or a mountain, is beautiful? The concept of beauty, and indeed all concepts of value and meaning, require observers. Without a mind to observe it, a waterfall is only a waterfall, a mountain is only a mountain. It is we conscious matter, the rarest of all forms of matter, that can take stock and record and announce this cosmic panorama of existence before us.

I realize that there is a certain amount of circularity in the above comments. For meaning is relevant, perhaps, only in the context of minds and intelligence. If the minds don't exist, then neither does meaning. However, the fact is that we do exist. And we have minds. We have thoughts. The physicists may contemplate billions of self-consistent universes that do not have planets or stars or living material, but we should not neglect our own modest universe and the fact of our own existence. And even though I have argued that our bodies and brains are nothing more than material atoms and molecules, we have created our own cosmos of meaning. We make societies. We create values. We make cities. We make science and art. And we have done so as far back as recorded history.

As I mentioned in the chapter "One Hundred Billion," the British philosopher Colin McGinn argues that it is impossible to understand the phenomenon of consciousness because we cannot get outside of our minds to discuss it. We are inescapably trapped within the network of neurons whose mysterious experience we attempt to analyze. Likewise, I would argue that we are imprisoned within our own cosmos of meaning. We cannot imagine a universe without meaning. I am not talking necessarily about some grand cosmic meaning, or a divine meaning, or even a lasting, eternal meaning—just the simple, particular meaning of everyday events, fleeting events like the momentary play of light on a lake, or the birth of a child. For better or for worse, meaning is part of the way we exist in the world.

And given our existence, our universe must have meaning, big and small meanings. I have not met any of the lifeforms living out there in the vast cosmos beyond Earth. But I would be astonished if none of them were intelligent, as I define intelligence. And I would be further astonished if those intelligences were not, like us, making science and art and attempting to take stock and record this cosmic panorama of existence. We share with those other beings not the mysterious, transcendent essence of vitalism, but the highly improbable fact of being alive.

INFINITY

INFINITY

Cosmic Biocentrism

In 1979, the late distinguished theoretical physicist Freeman Dyson indulged himself in one of the more daring speculations of the scientific imagination: the fate of the universe and intelligent life in the *extremely* remote future. We're not talking just hundreds of thousands of years in the future, when the next ice age occurs. Or even billions of years, when the Sun expands, turns into a "red giant" star, and incinerates the Earth. We're talking millions of billions of years, when all the stars in space have burned out and planets have been tossed from their solar systems by chance encounters with wandering stars. And even beyond. Somewhat surprisingly, physicist Dyson writes, "It is impossible to calculate in detail the long-range future of the universe without including the effects of life and intelligence." He then goes on to describe a scheme whereby intelligent life might survive in this grim future—by relocating consciousness and memory from flesh-and-blood bodies to large structures of particles, like floating clouds. To survive, these "intelligent structures" must go into long periods of hibernation between active periods of cautious nibbling at the dwindling supplies of energy.

Dyson's paper, brimming with mathematical computations, was published under the title "Time Without End: Physics and Biology in an Open Universe." The physicist was not unaware of the speculative nature of his predictions. Partly as cover, he quotes another great theoretical physicist, Steven Weinberg, who had recently published a book titled *The First Three Minutes*. Weinberg's book did for the very beginning of time what Dyson was hoping to do for the very end. "This is often the way it is in physics," wrote Weinberg. "Our mistake is not that we take our theories too seriously, but that we do not take them seriously enough."

Dyson, who died in February 2020 at age ninety-six, was a shy, small, elf-like man. Born in England to a musical composer father and lawyer mother, he demonstrated a high talent for mathematics at an early age. His older sister Alice remembers her little brother surrounded by encyclopedias and sheets of paper on which he was calculating things. In World War II, not quite twenty years old, Dyson was recruited by the RAF Bomber Command to compute the ideal pattern of bomber formations for the Royal Air Force. He studied mathematics at Trinity College, Cambridge, but never bothered to get a PhD. In 1947, Dyson moved to the United States and, a few years later, took up a permanent post at the famous Institute for Advanced Study in Princeton. Many physicists think that Dyson should have shared the 1965 Nobel Prize for his work in understanding how light interacts with matter, taking into account both quantum physics and Einstein's relativity.

Dyson was always a visionary. In the late 1950s, he led Project Orion, which proposed that spacecraft could be propelled by detonating a series of atomic bombs in their tail sections. A few years later, in 1960, he outlined something now called Dyson spheres, describing how an advanced civilization could utilize most of the energy of a star by building a light-collecting sphere around it. The Dyson tree is a genetically engineered plant, living in the open spaces in a comet and providing a sustainable atmosphere for human habitation.

Dyson's idea about the continuing survival of intelligent life in the infinite future was dubbed "Dyson's eternal intelligence." Like many of his other futuristic speculations, these new ideas were batted about in the scientific community, considered interesting and controversial and possibly useless, and grabbed up by science fiction writers.

A natural question about all such cosmic ruminations, of course, is: Are they just intellectual amusement, or do they tell us anything important about ourselves here and now, on planet Earth in the twenty-first century? Certainly, Copernicus's idea that the Sun, not the Earth, lies at the center of the solar system has had profound philosophical and theological impact in the here and now. Likewise, the recent discovery that a large fraction of stars have habitable planets, at the right distance from their central stars for liquid water.

Dyson's eternal intelligence percolated along with varying degrees of interest for twenty years. Then, in 1998, a new scientific discovery shook everything up. Astronomers found that the universe is not expanding at the leisurely rate accepted for over a half century and assumed in Dyson's calculations. Instead, the universe is expanding at an *accelerating* rate. That is, galaxies are flying away from one another at a speed increasing exponentially in time. As a result, in a mere hundred billion years, we and our local group of galaxies will be permanently cut off from the rest of the universe, as if we had fallen into a black hole. No light, energy, or anything else from the rest of the universe will ever reach us. (Our own Sun will have burned out long before this time, around ten billion years from now.) We will be imprisoned within a cell of limited size—large by Earthly standards but small in cosmic terms—preventing Dyson's intelligent structures from continually growing and storing information in ever larger domains. The night sky will become completely black, space will become colder and colder, and all remaining available energy will diminish to nothing. At some point thereafter, perhaps in another few hundred billion years, that will be the end of life—not just life like ours or even the life embodied by Dyson's "intelligent structures," but all life. Such a final demise will occur not only in our cosmic neck of the woods, but everywhere in the universe. The universe will continue to churn along forever, for infinite time, but the "era of life" will have passed. That result and its implications, I suggest, could be even more profound than Dyson's eternal intelligence.

In the previous chapter, I discussed the rarity of life in material terms. To contemplate its rarity in terms of the "era of life," one needs first to understand something of the enormous scales of distance and time in the universe. Seldom does daily existence afford any sense of our place in the cosmos, but eclipses provide a slight inkling. Like many Americans, I watched the solar eclipse of August 2017. My daughter and son-in-law and their two children were visiting my wife and me in coastal Maine, not the perfect location but a 58 percent eclipse according to the estimates. A couple of days before the event, we realized that we didn't have the proper equipment, so we began calling around to procure eclipse sunglasses. All the nearby stores were completely sold out. Eventually, my wife located a tiny library in a town called China, Maine, about an hour and a half's drive away, that had a good supply. She got a woman on the phone who said that they were closing soon, but she would leave a stock of eclipse sunglasses out in front of the library in a cooler, with a sign saying "Take only as many as you need." My wife made the trip.

Meanwhile, our four-year-old granddaughter got wind that something important was in the works and asked me to explain eclipses. I took out several pieces of fruit, one being the Earth, one the Moon, and one the Sun, and put them in front of her in an arrangement such that the fruit moon obscured the fruit sun. "Can you show me that on the computer?" she asked.

My son-in-law was not satisfied with the fruit analogue demonstration, or the computer simulation, or the real-time direct eclipse-sunglass experience. He got out a food colander and projected a hundred crescent Suns on the deck.

During the eclipse, as the light dimmed, the animals around us began behaving strangely. The birds' squawking didn't seem normal. The squirrels scampered in unnatural ways—or at least it seemed so to us. The monarch butterflies hovering about the chives in our garden swooped and fluttered as if in a trance. After a half hour of the eclipse, we'd had enough. We put down our sunglasses and colanders and got on with the rest of our day.

But something profound had happened to us. For a moment, we were aware of ourselves in the universe. We were aware of the cosmic nature of things, of the Moon as an enormous round ball orbiting the Earth, and the Earth as another ball orbiting the Sun and spinning about on its axis. And the immensity of space. Spectacular things are occurring out there, whether we notice or not.

My granddaughter asked me how far away the Sun is. That question I couldn't answer with apples and oranges. But if you traveled to the Sun on a high-speed train, say at two hundred miles per hour, it would take about fifty years. She nodded.

To get to the nearest star beyond the Sun on the same train would take about fifteen million years. It was Isaac Newton who first managed to estimate that distance, scribbling his calculations in a spidery script with a quill dipped in the ink of oak galls. (Only someone as extraordinary as Newton could have been the first to perform such a calculation and have it go almost unnoticed among his other achievements.) If one assumes that stars are similar things to our Sun, Newton asked, how far away would our Sun have to be in order to appear as faint as nearby stars? The challenge in such a calculation is how to compare the brightness of the Sun to that of a star. In the mid-seventeenth century, Newton didn't have electronic photocells at his disposal. He did, however, know that at certain times of the year the planet Saturn appears about as bright as a bright star. That planet glows because of reflected light from the Sun. Figuring out the fraction of the Sun's light that Saturn intercepts, Newton was able to get his answer: about thirty million million miles to the closest stars. His calculation, titled "On the Distance of the Stars," occupies only a single page in his masterwork, the *Principia*.

To deal with such huge distances and even larger ones, astronomers use a unit of distance called the light-year, the distance that light can travel in one year. In these terms, the nearest star, called Alpha Centauri, is about five light-years away. In other words, a light ray emitted from that star and traveling through space at 186,000 miles per second would take five years to reach the Earth.

Newton's estimate of this distance was far larger than any distance imagined in human history—fantastically larger than the circumference of the Earth or even the distance to the Sun (which the ancient Greeks had estimated). One should think of ants in an anthill in Cincinnati trying to imagine the distance to San Francisco.

But we're just getting started, astronomically speaking. When we look up on a dark and clear night, we see a beautiful white sash overhead. That's our galaxy, the Milky Way, a swarm of about a hundred billion stars. How to measure its size? For almost 250 years after Newton, no one knew. Then, in 1912, a mostly deaf astronomer named Henrietta Leavitt, working at the Harvard College Observatory, devised a completely new method to determine the distances to faraway stars. Certain stars, called Cepheid variables, were known to oscillate in brightness. Leavitt discovered that the cycle times of such stars are closely related to their intrinsic luminosities (wattages). More luminous stars have longer cycle times. Measure the cycle time of such a star and you know its intrinsic luminosity. Then, by comparing its intrinsic luminosity to how bright it *appears* in the sky, you can infer its distance, just as you can gauge the distance to an approaching car in the night if you know the wattage of its headlights. Cepheid variables are scattered throughout the cosmos and conveniently serve as cosmic distance signs in the highways of space. Leavitt, always referred to as "Miss Leavitt," received no honors and little recognition during her lifetime and is almost unknown outside the astronomical community.

In the 1920s, using Leavitt's results, astronomers were able to measure the size of the white sash of the Milky Way, which we now know to be a hundred thousand light-years across. At the time, there was a strong debate about other astronomical apparitions: Were the faint, nebulous smudges seen through telescopes part of our Milky Way, or something else? By identifying Cepheid variable stars in these smudges, the astronomer Edwin Hubble was able to determine that many of them are entire galaxies. The nearest large one, called Andromeda, is a couple of million light-years away. On average, each galaxy is separated from the nearest neighboring galaxy by ten or twenty galaxy diameters.

This picture of deep space was conceived by creatures roughly two meters in height, pondering the cosmos from one planet on the outskirts of one galaxy in the plentitude of galaxies. On January 22, 1926, *The New York Times* published a brief and blasé note about Hubble's discovery, under the headline "Another Universe Seen by Astronomer":

For many years, astronomers have speculated as to whether various nebulous formations in the heavens belong to this universe or were "island" universes of

their own, immeasurable distances away...Evidence
that another universe really exists is offered by Dr. Edwin Hubble in a study today published by the University of Chicago in the Astrophysical Journal. He found
that this external galaxy, similar in many ways to our
own, although entirely outside the Earth's galactic system, is 700,000 light-years away [an underestimate but
still much larger than the Milky Way].

Human beings have done better at contemplating very
long periods of time. In ancient Hinduism, the life span
of a deva (minor god) was thought to be about 10,000 deva
years, each deva year about 100 Earthly years, for a total of a
million years. One day in the life of Brahma, the Creator God,
was 1,000 deva lifetimes, estimated at about 4 billion years.
That long time unit was called a *kalpa*. Evidently, these succes-
sively longer time scales were arrived at simply by multiplying
the previous time scale by a few factors of ten, without any
knowledge of the physical world.

The Hindus believed in a cyclic universe. The life cycle of
the entire universe was thought to be a hundred years in the
life of Brahma, working out to be about three hundred thousand billion years. By accident, this time period is about the
time required for all stars to burn out.

Buddhists also used the kalpa as a unit of cosmic time,
but the Buddha demurred from specifying the length of the
kalpa in human years. However, he did give a vivid illustration:
Suppose you have a very large mountain 16 miles in height
and 16 miles in width. If you swipe the mountain with a piece

of silk once every hundred years, the mountain will be completely worn away before the end of a kalpa. (This conjecture has not been verified.)

The first very long time period determined scientifically and fairly accurately occurred in the 1920s, when geologists used the rate of disintegration of uranium and other radioactive elements to estimate that the Earth is several billion years old. Then, in the 1930s and 1940s, with the realization that our Sun and all stars are powered by nuclear fusion at their centers, astronomers and physicists estimated the age of our Sun: about five billion years.

In 1929, Edwin Hubble, again analyzing data from the giant telescope at Mount Wilson, California, discovered evidence that the universe is expanding—probably the most important cosmic discovery of all time. According to the Big Bang model and recent observations of the cosmos, we expect that the universe will keep expanding *forever*, growing colder and increasingly dilute. It was in such a context that Dyson considered how life could survive into the remote future, and perhaps forever.

We have now arrived at the point to reconsider the rarity of life, both in space and in time. One of the first people to propose the existence of life beyond planet Earth was the early Roman poet Lucretius (ca. 50 BC). In his great book *On the Nature of Things*, in which he espouses a purely materialist view of the cosmos in order to combat the supernatural power of the gods, Lucretius writes: "It is in the highest degree unlikely that this Earth and sky is the only one to have been created ... Nothing in the universe is the only one of its kind, unique and solitary in its birth and growth ... You are bound therefore to acknowledge that in other regions there are other Earths and various tribes of men and breeds of beasts."

As discussed in the last chapter, based on recent observations of the Kepler satellite, designed to search for "habitable" planets, we can estimate that the fraction of all material in the cosmos in living form is no larger than one-billionth of one-billionth. Life is indeed rare in space.

The discovery of cosmic acceleration makes life rare also in time. That is, only for a brief phase in the long history of the universe can life exist. "Brief," of course, is a relative term. Let me elaborate. Life and all complex structures require the larger atoms such as carbon and oxygen and nitrogen. (Even computers, which may someday be deemed alive, require heavy elements in the form of silicon.) The smallest atoms, hydrogen and helium, simply do not have enough structural components to build much of anything. We have a great deal of evidence that the larger atoms were made in the nuclear fusion reactions in stars. The first stars, in turn, could not form until the universe was about one billion years old, as they required a slow condensation and contraction of giant clouds of gas. So the beginning of the "era of life" was around one billion years after the Big Bang. At the other end, as previously discussed, life will probably not exist after the universe is about a thousand billion years old.

How can we think of this range of the era of life, from one billion years to a thousand billion years? In dealing with such large numbers and their ranges, it is most useful to think in terms of powers of ten. So here we have three powers of ten, from one billion to one thousand billion. What should we compare this range to? We cannot compare it to infinity, the duration of an infinitely expanding universe. No number can be compared to infinity. Failing that, we can compare it to the longest period of time where we believe some final qualitative change occurs. That would be the time in the remote future, long after the era of life, when all matter in the cosmos disintegrates, in a process called "proton decay." That era is estimated to be about a million billion billion billion (10^{33}) years from now. Beyond that remote period, no further conceivable changes would occur as the universe spins away into nothingness.

The *earliest* time when we have a fair understanding of the universe was about a millionth of a trillionth of a trillionth of a trillionth (10^{-42}) of a second after the Big Bang, about ten times the Planck era, discussed in the chapter "Between Nothingness and Infinity." From that earliest moment of understanding to the last moment of understanding, when all matter disintegrates, is a range of about 82 powers of ten. In sum, the evolving universe, as scientists believe they understand it, lasts for something like 82 powers of ten, while the era of life occupies only 3 powers of ten.

Evidently, life in our universe is a flash in the pan, a few moments in the vast unfolding of time and space in the cosmos. What are we to make of such a fact? For this writer, a realization of the scarcity of life makes me feel some ineffable connection to other living things, in a manner I've not experienced before. Perhaps it is mostly an intellectual connection, but not all of it. There is a kinship in being among those few grains of sand in the desert, or present during the relatively brief era of life in the vast temporal sprawl of the universe. Even though I will never contact or know other life beyond planet Earth, I am part of something rare and unique, never to pass this way again. Almost certainly, there are other thinking beings out there in the infinity of space who have their own astronomers and physicists and biologists (and painters and writers), and who have reached the same conclusion. We will likely never exchange a single word, but we have all realized the rarity of our existence and connection to one another. As mentioned in the last chapter, we are connected in that we are "fellow" observers of the cosmos. But we are also connected simply by our rarity, in time and in space. It is a thought too grand to fathom. But then again the knowledge that the atoms in our bodies were made in stars—a concept unanimously endorsed by the scientific community—is also difficult to fathom.

Early in the twentieth century, the Alsatian philosopher and polymath Albert Schweitzer introduced a concept he called *Ehrfurcht vor dem Leben,* which translates to "Reverence for Life." According to Schweitzer's autobiography, one day in 1915, while traveling on a river in Africa, the forty-year-old Schweitzer witnessed all at once the Sun shimmering on the water, the background of tropical forest, and a herd of hippopotamuses basking on the banks of the river. Suddenly he felt "the reverence for life." Later, Schweitzer put it this way: "I am life that wills to live in the midst of other life that wills to live."

Schweitzer's "Reverence for Life" became an underpinning of the more recent notion called "biocentrism"—a philosophical view that extends ethical value and connection to all living things. This view is explicitly nonanthropomorphic. Such an attitude is not new. It can be found in ancient religion and philosophy, including Buddhism. In modern times, the concept of biocentrism has been invoked by advocates of biodiversity, of protection of the environment, and of animal rights.

With the discoveries of the Kepler satellite in the last five years, it is almost certain that life exists elsewhere in the universe. (Given the unimaginable number of habitable planets, the absence of life beyond Earth would be like the absence of fires in a million dry forests, year after year after year.) The Kepler discoveries, plus the rarity of life in time and in space discussed here, lead to a concept I will call "cosmic biocentrism." By that, I mean that the rarity and preciousness of life provides a kinship to all living things in the universe. I cannot imagine what kinds of thoughts, what kinds of values and principles, other living beings might have. But we share something in the vast corridors of this cosmos we find ourselves in. What exactly is it we share? Certainly, the mundane attributes of "life": the ability to separate ourselves from our surroundings, to utilize energy sources, to grow, to reproduce, to evolve. I would argue that we "conscious" beings share something more during our relatively brief moment in the "era of life": the ability to witness and reflect on the spectacle of existence, a spectacle that is at once mysterious, joyous, tragic, trembling, majestic, confusing, comic, nurturing, unpredictable and predictable, ecstatic, beautiful, cruel, sacred, devastating, exhilarating. The cosmos will grind on for eternity long after we're gone, cold and unobserved. But for these few powers of ten, we have been. We have seen, we have felt, we have lived.

The Man Who Knows Infinity

In Jorge Luis Borges's story "The Book of Sand," a mysterious stranger knocks on the door of the narrator and offers to sell him a Bible he came by in a small village in India. The book shows the wear of many hands. The stranger says that the illiterate peasant who gave him the book called it the Book of Sand, "because neither sand nor this book has a beginning or an end." Opening the volume, the narrator finds that its pages are rumpled and badly set, with an unpredictable Arabic numeral in the upper corner of each page. The stranger suggests that the narrator try to find the first page. It is impossible. No matter how close to the beginning he explores, several pages always remain between the cover and his hand. "It was as though they grew from the very book." The stranger then asks the narrator to find the end of the book. Again, he fails. "It can't be," says the narrator. "It can't be, but it is," says the Bible peddler. "The number of pages in this book is literally infinite. No page is the first page; no page is the last." The stranger pauses and reflects. "If space is infinite, we are anywhere, at any point in space. If time is infinite, we are at any point in time." (Note to the observant reader: We cannot be at any point in time. Life can exist only during a relatively short period of cosmic history, as discussed in the last chapter.)

Thoughts of the infinite have mesmerized and confounded human beings through the millennia. For mathematicians, infinity is an intellectual playground, where an endless string of fractions can add up to 1. For astronomers, the question is whether outer space goes on and on and on and on ad infinitum. And if it does, as cosmologists now believe, unsettling consequences abound. For one, there should be an infinite number of copies of each of us somewhere out there in the cosmos. Because even a situation of minuscule probability—like the creation of a particular individual's exact arrangement of atoms—when multiplied by an infinite number of trials, repeats itself an infinite number of times. Infinity multiplied by any number (except 0) equals infinity.

Measurements of infinity are impossible, or at least impossible according to the usual notions of size. If you cut infinity in half, each half is still infinite. If a weary traveler arrives at a fully occupied hotel of infinite size, no problem. You simply move the guest in room 1 into room 2, the guest in room 2 into room 3, and so on ad infinitum. In the process, you've accommodated all of the previous guests and freed up room 1 for the new arrival. There's always room at the infinity hotel.

We can play games with infinity, but we cannot *visualize* infinity. By contrast, we can visualize flying horses. We've seen horses, and we've seen birds, so we can mentally implant wings on a horse and send it aloft. Not so with infinity. The unvisu-alizability of infinity is part of its mystique.

The first recorded conception of infinity seems to have occurred around 600 BC and is attributed to the Greek phi-losopher Anaximander, who used the word *apeiron*, meaning "unbounded" or "limitless." For Anaximander, the Earth and the heavens and all material things were caused by the infinite, although infinity itself was not a material substance. Other ancient Greek philosophers held that infinity was a negative, even an evil, because the inability to measure a thing was con-sidered a shortcoming of the thing—with the exception of the infinite and immeasurable One. About the same time as Anaximander, the Chinese employed the word *wuji*, meaning "boundless," and *wuqiong*, meaning "endless," and believed that the infinite was very close to nothingness. (An inter-esting perspective on Pascal's ideas, discussed in "Between Nothingness and Infinity.") In Chinese thought, being and nonbeing, like yin and yang, are in harmony with each other—thus the kinship of infinity and nothingness. A few centuries later, Aristotle argued that infinity does not actually exist. He conceded something he called *potential infinity*, such as the whole numbers. For any number, you can always create a big-ger number by adding one to it. This process can continue as long as your stamina holds out, but you can never get to infinity.

Indeed, one of the many intriguing properties of infinity is that you can't get there from here. Infinity is not simply more and more of the finite. It seems to be of a completely differ-ent nature, although pieces of it may appear finite, like large numbers, or like large volumes of space. Infinity is a thing unto itself. All we see and experience has limits, boundaries, tangibilities. Not so with infinity. For similar reasons, St. Augustine, Spinoza, and other theological thinkers have associated infinity with God: the unlimited power of God, the unlimited knowledge of God, the unboundedness of God. "God is everywhere, and in all things, inasmuch as He is boundless and infinite," said St. Thomas Aquinas.

Beyond the religious sphere of the immaterial world, physicists believe that there may be infinite things in the *material world* as well. But this belief can never be proven. You can't get there from here.

Most of us have our first glimmerings of infinity as children, when we look up at the night sky for the first time. Or when we go to sea, out of sight of land, and gaze upon the ocean extending on and on until it meets the horizon. But these are only glimmerings, like counting to a few thousand in Aristotle's potential infinity. We're overwhelmed. But we haven't come close.

The concept of infinity remains a controversial and paradoxical topic today, galvanizing international conferences and heated scholarly disputes. Can physical forces ever be infinite in strength? Can physical space extend beyond galaxy after galaxy without limit? Is there an infinity between the infinity of the whole numbers and the infinity of all numbers? In May 2013, a panel of scientists and mathematicians gathered in New York City to discuss the profound conundrums surrounding infinity. William Hugh Woodin, a mathematician at the University of California, Berkeley, put it this way: "It's kind of like mathematics lives on a stable island—we've built a solid foundation. Then, there's the wild land out there. That's infinity."

The person on planet Earth who may have come up with the most expansive conception of spatial infinity is the theoretical physicist Andrei Linde, a professor at Stanford University. Professor Linde works only with pencil and paper. Now seventy-two years old, he was born and grew up in Moscow and received his PhD in physics there from the Lebedev Physical Institute. Both of his parents were physicists. He married a physicist, Renata Kallosh (also a professor at Stanford). In 1990, Linde and Kallosh moved to the United States, where he took up his current academic position.

In the early 1980s, Linde proposed a radical theory of the origin of the universe. Linde's theory, a revision of MIT physicist Alan Guth's 1981 theory, itself a revision of the 1927 Big Bang model, is called "eternal chaotic inflation." The theory posits that in its infancy, our universe went through a period of highly rapid expansion, much faster than in the standard Big Bang model. In a tiny fraction of a second, a region of space smaller than an atom "inflated" to a size large enough to encompass all of the matter and energy that we can see today. That much of the inflation theory was articulated in Guth's paper. Linde's theory goes further. It predicts that our universe is necessarily one of a vast number of universes, each of which is constantly and randomly spawning new universes in an unending chain of cosmic creation, extending into the future for eternity. Some of those universes, and perhaps our own, should be infinite in extent. In our particular universe, the period of highly rapid expansion would have been completed and done with when our universe was 0.000000000000 00000000000000000001 seconds old.

It would be easy to dismiss such speculations as science fiction. But the fantastic speculations of scientists have often found a grip on reality. Two hundred years ago, who would have thought we would be able to decipher the microscopic chemical code that creates living organisms and to alter that code as if rearranging a deck of cards? Or build tiny boxes that could communicate pictures and voices through space? Linde's speculations are backed up by serious equations, and a number of important predictions of the Guth-Linde inflation theory (but not the existence of infinity) have been confirmed by experiment.

In the scientific community, Linde is widely regarded as a physicist of the first rank. He has won most of the major prizes in physics except for the Nobel. To name a few: the Dirac medal of the International Center for Theoretical Physics in Trieste (shared with Guth and Paul Steinhardt); the Gruber Prize in cosmology (shared with Guth); the Humboldt Prize in Germany; the Kavli Prize of the Norwegian Academy of Sciences and the Kavli Foundation (shared with Guth and Alexei Starobinsky); the Medal of the Institute of Astrophysics in Paris; and, in 2012, the inaugural Fundamental Physics Prize (shared with Guth and others), which carries an award of $3 million per person, more than twice the prize money of the Nobel.

Linde does not have a small opinion of himself. When I met him the first time, in 1987, a few years after his most important work on the inflation theory, he told me about his discovery with these words: "I easily understood what Guth was trying to do. But I did not understand *how* it [inflation] could be done, since we have seen that the inhomogeneities [in Guth's original theory] were large [contradicting observations]. I just had the feeling that it was impossible for God not to use such a good possibility to simplify His work, the creation of the universe...I was simultaneously discussing similar matters with Rubakov [by telephone]...I was sitting in my bathroom, since all my children and my wife were already sleeping at the time...After the whole picture had crystallized, I was very excited. I came to my wife and I woke her up and I said: 'It seems that I know how the universe originated.'"

I visited Linde recently at his home in Stanford, California, to get an update on his theory and its place in our view of the world. Linde and his wife live in a lush neighborhood of winding streets, tropical gardens, and houses set up on hills. He was casually dressed in a black fleece sweater over a black T-shirt, black pants, and sandals with black socks—all in dramatic contrast with his snowy white hair. His English is good but retains a thick Russian accent. We sat at the kitchen table. On the wall, a clock, a map of Tuscany, painted ceramic jars on a shelf. His wife prepared a delicious lunch of tortellini and salad.

First off, I asked Professor Linde if he believes that spatial infinity truly exists. "Do you think dinosaurs truly existed?" he replied and paused. "Everything works *as if* spatial infinity exists." Linde is careful with language. He distinguishes between reality, which we can never know, and our models and inferences about reality. Linde has always had a strong interest in philosophy. He remembers having debates with high school classmates about science versus art. One of his teenaged philosophical ideas, and an idea that he has not completely abandoned, was that "feelings" are actual objects. Young Linde theorized that when two people are communicating, verbally or nonverbally, their feeling-objects are shared simultaneously. However, in his science classes he learned that Einstein's theory of relativity forbids any communication faster than the speed of light. He decided that he had better study physics first, so as not to make such "mistakes."

I asked Professor Linde how he thought about infinity, whether he attempted to visualize it. "No matter how far you go, you can go farther," he said. Then he made an analogy to a garden. "But there's no fence." A week earlier, Robert Jaffe, a theoretical physicist at MIT, told me that he didn't find "the concept of infinity as troubling as the concomitant concept of nothingness." Linde said that he would not be particularly bothered by having many copies of himself in outer space in an infinite universe. However, he did allow that it would be "profoundly significant if their thoughts were the same as mine."

Anaximander's conception of infinity was abstract and could not reasonably be associated with physical space. In fact, the early Greek philosophers pictured the cosmos as limited in size, with an outer boundary, although the actual distances were not known. In this picture, often associated with Aristotle, the Earth resided at the center of a set of concentric spheres. Going outward was the sphere of the Moon, then Mercury, Venus, the Sun, Mars, Jupiter, and Saturn. Beyond the sphere of Saturn was the "sphere of the fixed stars," with each star attached to this shell like a bulb on a spherical Christmas tree. And beyond this stellar sphere was the last and outermost sphere, the "primum mobile" ("first moved"), spun by the finger of God. In the sixteenth century, Copernicus changed almost everything by putting the Sun at the center of the solar system. But the Polish scientist left unchanged the idea of a finite universe. The unquenchable stars still dangled from an outermost sphere.

The first person to postulate in concrete terms a spatially infinite universe seems to have been an English mathematician and astronomer named Thomas Digges (1546–1595). In 1576, Digges published a new edition of his father's perpetual almanac, *A Prognostication Everlasting.* With the elder Digges long departed, his son was emboldened to add an unauthorized appendix titled "A Perfit Description of the Caelestiall Orbes according to the most aunciente doctrine of the Pythagoreans, latelye revived by Copernicus and by Geometricall Demonstrations approved." In that appendix, Digges abolished the sphere of the stars. At the center of his diagram is the face of the Sun, with spiky rays issuing forth. Then the "orbes" of the planets. And beyond this region and extending to the edge of the page are the stars, scattered here and there through infinite space.

About one thing Digges and Copernicus and Aristotle concurred. The cosmos on the whole was at rest—a magnificent and immortal cathedral. The cosmos had existed forever and would exist forever, from the infinite past to the infinite future. This peaceful conception sat quietly for another three hundred years. Even Einstein's cosmological model of 1917, based on his new theory of gravity, proposed a static and eternal universe.

Then came the Big Bang. In 1927, a Belgian priest and physicist named George Lemaître suggested that the previously observed outward motion of galaxies meant that the universe was expanding. Two years later, Lemaître's suggestion was confirmed by American astronomer Edwin Hubble, who found that the speed at which other galaxies are flying away from us is proportional to their distance—exactly the result as if all the galaxies were dots painted on an expanding balloon. From the viewpoint of *any* dot (galaxy), it appears that all the other dots are moving away from it. No dot is the center.

By measuring the rate at which the universe is expanding today, we can estimate when the universe "began," about fourteen billion years ago. Since that moment, the universe has been expanding, thinning out, and cooling. It is important to note that the balloon analogy is only an analogy. In particular, unlike a balloon, the universe could be infinite in extent. What astronomers mean by the statement that the universe is expanding is that the distance between any two galaxies is increasing with time.

The Big Bang model is more than an idea. It is a detailed set of equations describing how the universe has evolved since t = 0, specifying in quantitative detail such things as the average density and temperature of the universe at each point of time. The model has subsequently been supported by several pieces of evidence. For one, the age of the universe as calculated by its rate of expansion approximately agrees with the age of the oldest stars, calculated by our understanding of the physics of stars. For another, the Big Bang model predicts that there should be a flood of radio waves coming from all directions in outer space, produced when the universe was about 300,000 years old and now having a temperature of about 270 degrees below zero Celsius. That predicted flood of radio waves, called the cosmic background radiation, was discovered in 1965. There are other confirmed predictions as well, such as the observed proportions of the lightest chemical elements. The Big Bang theory does not say whether space and time existed before the cosmic balloon began expanding. That profound question would be left to Linde and others. (See the earlier chapter "What Came Before the Big Bang?")

Linde would have first heard about the Big Bang model as a university physics student in Moscow in the late 1960s. However, he was trained not as a cosmologist but as a particle physicist, as was Alan Guth. Particle physicists study nature at the smallest sizes, while cosmologists study it at the largest. The two branches of physics seemingly had little to do with each other. But in the early 1970s, Linde became interested in certain phenomena that occur at extremely high temperatures, far beyond what can be created in the laboratory, temperatures that could have existed only in the fantastically hot conditions of the infant universe. Describing one of his theories at this time, a prelude to his work on inflation, Linde said, "At the first glance, this theory seemed to be too exotic. We developed it in 1972, but for two years nobody believed us. People were laughing…" But in 1974, some American physicists confirmed the main conclusions.

This response to Linde's early work—first doubts and then often acceptance—seems to have been a pattern in his career.

In our conversation, we talked about the manner in which scientific theories, and especially maverick theories, are confronted by the scientific community. Linde described what he calls a strong "sociological" effect: the biases and prejudices of scientists, their institutional stature, and, of course, the inherent caution of the scientific enterprise. Linde himself is not a cautious person. His colleagues describe him as someone who shoots from the hip with lots of ideas, some right, some wrong, a person of extreme self-confidence, a showman in his popular lectures and articles.

By the early 1970s some physicists were worrying about problems with the Big Bang model, despite its successes. One troubling concern, for example, is that the cosmic radio waves are highly uniform in temperature, no matter what direction we look. There are two possible explanations for this result: either the universe began in an extremely uniform condition, with all parts at the same temperature, or any initial non-uniformities were smoothed out in time, much as hot and cold water in a bathtub will come to the same temperature by exchanging heat. However, heat exchange takes time. According to the Big Bang model, the far-flung distant parts of the universe we see today would not have had time to exchange heat during the first 300,000 years of the universe, when the cosmic radio waves were created. Thus, the second explanation doesn't work. On the other hand, the first explanation is considered unpalatable because it sweeps the problem under the rug, saying: "It is what it is because it was what it was." In general, physicists detest such arguments. They prefer to explain everything in the physical universe as the necessary consequence of a few calculable laws and principles rather than as "accidents" of initial conditions, beyond their ability to calculate.

The Guth-Linde inflation theory solves the puzzle of the cosmic radio waves as well as other problems with the Big Bang model. During the period in the infant universe when space was expanding at blinding speed, a very tiny patch of space, tiny enough that all its parts could have homogenized, would have quickly inflated to encompass today's entire observable universe. No matter what the initial conditions, inflation would have produced a universe of uniform temperature.

Most importantly, the inflation theory explains *why* such inflation would occur and includes equations for the various energies and forces involved. The key ingredient of the theory, and the cause of the extremely rapid expansion of the infant cosmos, is a kind of energy called a *scalar field*. Most energy fields, like gravity, are invisible, yet they can exert forces. Some scalar fields produce a repulsive gravitational force. They push things apart rather than pull things together.

What I call the Guth-Linde theory was developed over a period of several years, from 1979 to 1986, beginning with work by Alexei Starobinsky in Moscow. During that period, there were various versions of the theory, problems arising and fixed, new ideas proposed, and an assortment of other physicists involved.

One of Linde's ideas is that in the early universe, scalar field energy should be constantly created at various magnitudes, due to quantum effects. A strange aspect of quantum physics is that energy and matter can suddenly appear out of nothing for short periods of time. If you could examine space with a strong enough microscope, you would find that it is constantly fluctuating, seething with ghost-like particles and energies that randomly appear and disappear. Quantum phenomena are normally apparent only in the tiny world of the atom, but near $t = 0$ the entire observable universe was smaller than an atom. If at a certain point in the infant universe sufficient scalar field energy materialized, its repulsive gravitational effect would cause space to expand so rapidly that an entire universe would be created. Since such quantum fluctuations would be going on at random places and times (the "chaos" in Linde's eternal chaotic inflation theory), new universes would be constantly forming.

Indeed, Linde's theory requires that we redefine what we mean by "universe." Some physicists now take the word to mean a region of space that will be quarantined into the infinite future. This region may have been in contact with other parts of the cosmos in the past but can never communicate with the rest of the cosmos in the future. For all practical purposes, each such region is its own universe. In the mind-bending way in which the geometry of space is altered by gravity according to Einstein, it is possible that there be multiple universes, each infinite in extent. The new universes created by quantum fluctuations are predicted to have a wide range of properties—some might be infinite in extent, others finite; some might have the right conditions to make stars and planets and life; others might be lifeless and unformed deserts of subatomic particles and energy; some might even have different dimensions than our own universe. In this vision, there would be an endless creation of new universes, each with its own Big Bang beginning. Our t = 0 would not be the beginning of space and time in the larger cosmos, only for our particular universe. In Linde's vision of reality, although everything in our universe passes away, the constellation of universes, continually spawning new universes, would represent a kind of immortality.

In some of his papers, Linde illustrates his eternal chaotic inflation model as a thick hedge of branching bulbs, each bulb a separate universe, connecting to ancestor bulbs and descendant bulbs by thin tubes. The entire collection of universes might be called the "cosmos." Sometimes, it's called the "multiverse." It is startling to look at Linde's picture and to realize that each bulb represents an entire universe, some containing stars and planets, cities, office buildings, trees, ants or antlike creatures, sunsets. Unfathomable—yet a human mind has fathomed this thick hedge of the imagination. "It can't be, but it is," says the Bible peddler in "The Book of Sand."

One cannot resist comparing Andrei Linde's "map of the universes" to the Babylonian Map of the World, one of the oldest known maps ever drawn by human beings, found on a stone tablet in what is present-day Iraq and now housed in the British Museum. In this ancient map of the known world (ca. 600 BC) the city of Babylon is perched on the Euphrates River, flowing north and south. Pictured and named (in Sanskrit) are a few other cities, including Uratu, Susa, Assyria, and Habban; a mountain; and a circular ocean (labeled as "bitter river") enveloping the inhabited cities. Finally, some unnamed and unknown outer regions represented by spikes radiating out from the circular ocean. Could one compare these unnamed spikes to the unnamed bulbs in Linde's map? Both lie far beyond the realm of physical exploration. Both require leaps of the imagination. Yet Linde's bulbs follow as logical consequences of certain mathematical equations. As Linde would acknowledge, those equations are also works of the human imagination, models of reality instead of reality itself. Linde's ideas are at once visionary and grounded in logical thinking. Although mathematically proficient in the manner of all theoretical physicists, Linde described himself to me as more intuitive than technical, a Steve Jobs more than a Steve Wozniak.

The Babylonian Map of the World is a static picture. By contrast, Linde's Map of the Universes suggests evolution and change, movement. The various universes spawn one another in time. A better comparison, then, might be found in Hindu cosmology, in which our universe is one of an infinite number of cycling universes. The totality has no beginning or end. The concept is described in the Bhagavata Purana:

> Every universe is covered by seven layers—earth, water, fire, air, sky, the total energy and false ego—each ten times greater than the previous one. There are innumerable universes besides this one, and although they are unlimitedly large, they move about like atoms in You. Therefore You are called unlimited.

I do not feel unlimited looking at Linde's Map of the Universes. Instead, I feel small and insignificant, like the Bible peddler who says that if space is infinite, we are nowhere in space, nowhere in time. How can anything we do be of consequence when we are nowhere in space, nowhere in time, when our brief lives are lived out on one small planet, itself one of zillions of planets in a universe that may be infinite in size, and our entire universe simply one bulb in Linde's thick hedge of universes? On the other hand, there may be something majestic in being a part, even a tiny part, of this unfathomable chain of being, this infinitude of existence. We pass away, our Sun will burn out, our universe may become a dark and lifeless void a hundred billion years from now— but, according to Linde, other universes are constantly being born, some surely with life, renewing something precious that cannot be named.

It is unlikely that we will ever know if Linde's infinity of universes exists. But the rest of the Guth-Linde inflation theory is being actively tested today. One of the most important tests, Linde explained to me, is a search for something called "B-mode polarization," a slight twisting pattern in the vibrations of the cosmic radio waves predicted by the inflation theory. A few years ago, astronomers thought they had discovered the phenomenon, an experimental confirmation that would probably have brought Nobels to Linde and Guth. On the morning of Thursday, March 6, 2014, a professor of astrophysics at Stanford named Chao-Lin Kuo knocked on the door of Linde's home. Dr. Kuo was accompanied by a camera crew. (The resulting video, made by Stanford University, was posted on YouTube eleven days later and has received over three million hits.) When they open the door, Linde and his wife appear stunned at the news. Renata gives Chao-Lin a big hug. Then the cameras follow Linde and Kuo as they go into the kitchen and share a bottle of champagne. We hear the pop of the cork. We see the clock on the wall, the map of Tuscany, the painted ceramic jars on a shelf. "We are talking about a billionth of a billionth of a billionth a millionth of a second after the Big Bang," says Linde. "Finally, it arrived," he says with a smile on his face. Eleven days after Dr. Kuo's visit, headlines appeared all over the world. In a *New York Times* article titled "Space Ripples Reveal Big Bang Smoking Gun," cosmologist Marc Kamionkowski of Johns Hopkins University said, "This is huge, as big as it gets." Max Tegmark, a cosmologist at MIT, said, "I think that if this stays true, it will go down as one of the greatest discoveries in the history of science." It didn't stay true. Or rather, the experimental results were correct, but misinterpreted. A follow-up analysis showed that the twisting effect was likely caused by ordinary dust in outer space rather than by the extremely exotic processes predicted by the Guth-Linde inflation theory. That realization did not deflate the theory, but it did leave more work to be done.

Refined measurements of the B-mode polarization, able to distinguish between mundane dust in the Milky Way and cosmic inflation in the infant universe, are now being conducted by the "Polar Bear" experiment in the Atacama Desert of northern Chile and by the "BICEP" experiment at the South Pole, among others. These experiments are international collaborations, including over a dozen institutions in the United States, England, Wales, France, and Canada. Thousands of scientists worldwide, both theorists and experimentalists, are actively working to test the inflation theory and to probe its consequences. Almost all cosmologists today accept it as the best working hypothesis we have of the first moments of our universe. The theory must be considered a triumph of the human mind.

Yet, Andrei Linde does not appear to be a man completely at peace with his place in the world. Something eludes him. When he talks about the history of the inflation theory, he seems to be still defending his ideas against naysayers and rival theorists, still competing with Guth and others for priority of discovery, still infused with a powerful desire for vindication. In my conversations with him and in his review articles and autobiographical statements, he portrays himself as someone who heroically developed a new view of the cosmos, struggled against doubters, corrected other people's mistakes and misunderstandings, and was often misunderstood himself. One story he enjoys telling is about a lecture that Stephen Hawking gave at the Sternberg Astronomical Institute in Moscow in October 1981. Linde was asked to translate for the Russian audience. At this time, various physicists, including Hawking, were trying to patch up a serious problem (too much inhomogeneity) in Guth's original inflation theory. Linde had devised his own inflation theory, a revision of Guth's theory, but it was not yet published. During the lecture, Hawking would mumble a few incoherent words, one of his graduate students familiar with his speech would translate into understandable English, and then Linde would translate into Russian. With this painfully slow process under way, Hawking announced that Linde had a good idea, but that it was wrong. Then for the next half hour, sitting in his wheelchair, he proceeded to explain why it was wrong. All the while, Linde had to translate. At the end of the lecture, Linde told the audience, "I have translated, but I disagree." He then took Hawking in his wheelchair to another room of the building, closed the door, and explained to him more details of his new theory. Evidently, Hawking had to admit that Linde was right after all. According to Linde, Hawking "was sitting there about one hour and a half and saying to me the same words: 'But you did not tell this before. But you did not tell this before.'"

Perhaps Linde's ego and bravado were essential for the conception of his phantasmagoric cosmology. Other scientists with equal brainpower but more cautious dispositions have not ventured nearly so far in their theories of the world. The equations are the equations, but they must be imagined and interpreted in the human mind, a particular human mind, a complex universe itself, endlessly variable in its quirks and possibilities.

"At the beginning, I was like a young kid, making discoveries," Linde told me. "Now I feel a deep responsibility. There are hundreds of people working on the theory of inflation and lots [of expensive] experiments to test it. You feel yourself a bit heavy with responsibility ... I would hate to die just being a physicist. I enjoy photography. That allows me to feel another part of my brain. There is something beyond physics that is not measurable ... Photography is my art. You need to have a first priority and then a second priority. When I was sixty, someone gave me a camera. With a camera, you can produce beauty. I can produce things that are better than what I see in museums. You see, I am now talking like an arrogant American. I am producing images that make my heart sing—both my photographs and the computer graphics illustrating inflation. I am among the first to see the beauty in it. Without the part of my mind beyond physics, I would be unable to create the computer graphics of cosmology."

Linde went to his computer and eagerly showed me his Flickr website, where he has posted hundreds of his photographs. "Sit down," he said and offered me a seat near the screen. One of his photographs, titled *Alcazar Dreams,* depicts a subterranean pool beneath the Patio del Crucero in Seville, Spain. He explained that the pool was built by the master of the castle for his lady friend. A series of stone arches, glowing in eerie orange light, bend over the elongated pool, one after another after another out to a distant vanishing point. Another image, titled *Hide Thy Face,* is an extreme close-up of the interior of an orchid. Around the outside edges unfolds a diaphanous blue halo. At the middle of the flower is a two-chambered yellow heart covered with red speckles, with white-and-red-striped arms emerging from it, and farther out pale green and yellow petals. Altogether, an intricate jewel, a tiny splash in infinity.

Notes

BETWEEN NOTHINGNESS AND INFINITY

7 "The whole visible world": Blaise Pascal, *Pensées* (Thoughts), translated by W. F. Trotter, Harvard Classics (New York: P. F. Collier & Son, 1909), vol. 48, pp. 27–28.

9 A famous portrait of Pascal: A good biography of Pascal is Marvin R. O'Connell's *Blaise Pascal* (Grand Rapids, MI: William B. Eerdmans, 1997).

11 "a man of the world among ascetics": T. S. Eliot, *Selected Essays* (London: Faber and Faber, 1931), pp. 411–12.

11 "sickness is the natural state": See, for example, Will Durant, *The Story of Civilization: Our Oriental Heritage* (New York: Simon and Schuster, 1935), chapter 2.

17 How many times should the size: Counted in doublings, human beings are halfway between an atom and a star. An atom is about 10^{-8} centimeters in size, a human being is about 10^2 centimeters in size, and a star is about 10^{11} centimeters in size.

18 "the rest / From Man or Angel": *Paradise Lost*, book 8, lines 71–75.

18 "The most beautiful experience we can have": Albert Einstein, in

"The World as I See It," originally published in *Forum and Century* 84 (1931): 193–94. It can be found in Albert Einstein, *Ideas and Opinions* (New York: Modern Library, 1994), p. 11.

WHAT CAME BEFORE THE BIG BANG?

23 On Wednesday, February 11, 1931: Einstein's attitudes toward a nonstatic cosmology and his trip to the Mount Wilson Observatory on February 11, 1931, are described in great detail in an article by historian Harry Nussbaumer, "Einstein's Conversion from His Static to an Expanding Universe," *European Physical Journal H* 39 (2014): 37–62.

23 "Photographers lunged at me": Einstein's diary entry of December 11, 1930 (*Albert Einstein Archives*, Amerika-Reise 1930, Archivnummer 29–134), translated ibid., p. 44.

23 Einstein dismissed the evolving cosmology: The Russian cosmologist was Alexander Friedmann, the Belgian Georges Lemaître.

24 "abominable": In Georges Lemaître's memories of talking to Einstein at the 1927 Solvay conference, "Rencontres avec A. Einstein," *Revue des Questions Scientifiques* 129 (1958).

24 "has smashed my old construction": *New York Times*, February 12, 1931, p. 15.

25 Physicists believe that in this quantum era: During the quantum era, the typical length scale was the "Planck length," 10^{-33} centimeters, and the typical temperature was the "Planck temperature," 10^{32} degrees.

26 "high risk, high gain": Carroll interview with AL, August 4, 2015.

26 "What If Time Really Exists?": Sean Carroll, http://arxiv.org/abs/0811.3772.

27 "I strongly believe that the low entropy": Carroll interview with AL, September 17, 2015.

30 "the universe was neither created nor destroyed": Stephen Hawking, *A Brief History of Time* (New York: Bantam Books, 1988), p. 136.

30 "The view from my previous office was better": This and other quotes from Vilenkin come from Vilenkin interview with AL, July 7, 2015.

32 "It is a mystery to me": Hartle interview with AL, July 29, 2015.

33 "What place, then, for a Creator?": Hawking, *Brief History of Time*, p. 141.

33 "As a Christian, I think there is a Being outside": Page interview with AL, September 11, 2015.

33 "One might think that adding the hypothesis": Don Page, "Guest Post: Don Page on God and Cosmology," in Sean Carroll's blog, *The Preposterous Universe*, March 20, 2015, http://www .preposterousuniverse.com/blog/2015/03/20/guest-post-don -page-on-god-and-cosmology/.

34 "In everyday life we talk about cause and effect": Carroll interview with AL, August 4, 2015.

34 "Causality within the universe": Page interview with AL, September 11, 2015.

ON NOTHINGNESS

37 "Nothing will come of nothing": *King Lear*, act I, scene 1.

37 "Man is equally incapable": Pascal, "The Misery of Man Without God," *Pensées*, section 72.

37 "The ... 'lumniferous ether' will prove": Original paper in German in *Annalen der Physik* 17 (1905): 891–921, translated by W. Perrett and G. R. Jeffrey in *The Principle of Relativity* (New York: Dover, 1952).

38 Following Aristotle, let me say: *The Categories*, translated by H. P. Cooke (Cambridge, MA: Harvard University Press, 1980).

45 "The first principle": "Cargo Cult Science" (adapted from a 1974 Caltech commencement address), the last chapter in Richard Feynman, *Surely You're Joking, Mr. Feynman!* (New York: Norton, 1985).

ATOMS

49 "It seems probable to me that God in the beginning": Isaac Newton, *Optics,* book III, part 1, translated by Andrew Motte and revised by Florian Cajori, in Encyclopaedia Britannica's *Great Books of the Western World* (Chicago: University of Chicago Press, 1987), vol. 34, p. 541.

50 *Pleasing substances are made of:* Paraphrase of Lucretius, *De Rerum Natura,* book 2, 398–407. See, for example, *De Rerum Natura,* translated by W. H. D. Rouse in Loeb Classical Library (Cambridge, MA: Harvard University Press, 1982), p. 127.

51 At an engaging internet site hosted: http://history.aip.org/history/exhibits/electron/jjsound.htm.

53 "As history unveiled itself in the new order": Henry Adams, "The Grammar of Science," in *The Education of Henry Adams* (1903: Boston: Houghton Mifflin, 1918), p. 458.

54 "It was quite the most incredible event": Ernest Rutherford in *Background to Modern Science,* ed. Joseph Needham and Walter Pagel (Cambridge: Cambridge University Press, 1938), p. 68.

56 "I could be surprised": Jerry Friedman interview with AL, May 28, 2004.

58 we cannot divide space: See Lee Smolin, "Atoms of Space and Time," *Scientific American,* January 2004.

MODERN PROMETHEUS

59 "I am by birth a Genevese": Mary Shelley, *Frankenstein; or, The Modern Prometheus* (1818), chapter 1.

59 "One of the phenomena": Ibid., chapter 4.

60 *Omne vivum ex vivo:* René Dubos, *Louis Pasteur: Free Lance of Science* (Cambridge, MA: Da Capo Press, 1960), p. 187.

62 In the 1950s, chemists showed that electrical discharges: This refers to the Miller-Urey experiment, in 1953.

62　The first creation of a synthetic cell: This refers to the work of Thomas Chang at McGill University.

62　The first hybrid gene: The first creation of an *altered* form of life, achieved by splicing together the genes of two different organisms, occurred in the early 1970s. This refers to the work of Paul Berg and colleagues in 1972.

62　The first synthesis of a complete set of genes: This refers to the work of J. Craig Venter and colleagues in 2010.

64　"The experiments I conducted there": Szostak, Nobel autobiography (2009), https://www.nobelprize.org/prizes/medicine/2009/szostak/biographical/.

64　"My father was often unhappy": Ibid.

65　"I had a growing feeling that my work in yeast": Ibid.

65　In a 2001 paper in the prominent journal: Jack W. Szostak, David P. Bartel, and P. Luigi Luisi, "Synthesizing Life," *Nature* 408 (January 18, 2001): 387.

66　"Although I have had some degree of success": Szostak, Nobel autobiography.

66　"One of the delights of the world of science": Ibid.

66　"a brilliant and energetic student": Ibid.

66　"I had the good fortune": Ibid.

67　"People get all tied up in knots": Jack Szostak interview with AL, Massachusetts General Hospital, July 17, 2019. All remaining quotes from Szostak come from this interview, unless otherwise indicated.

68　In 2003, Szostak and his colleagues demonstrated: "Experimental Models of Primitive Cellular Compartments: Encapsulation, Growth, and Division," *Science* 302 (October 24, 2003).

68　*The New York Times* published an article: Nicholas Wade, "How Did Life Begin?" *New York Times*, November 11, 2003.

68　*Scientific American* published an article: Sarah Graham, "Clay Could Have Encouraged First Cells to Form," October 24,

2003, https://www.scientificamerican.com/article/clay-could
-have-encourage/.

71 "In my view, the critical problem right now": Email to AL from
    Jack Szostak, August 5, 2019.

71 "Our approach has a ways to go": Ibid.

73 "Soul is the life force": Email to AL from Micah Greenstein,
    May 24, 2019.

73 "the rest from Man or Angel": Lines from *Paradise Lost*, book 8,
    lines 71–75.

74 "pried open one of the most forbidden": Michael Specter with
    Gina Kolata, "After Decades of Missteps, How Cloning Suc-
    ceeded," *New York Times*, March 3, 1997.

74 according to a recent Gallup survey: https://news.gallup.com/
    poll/6028/cloning.aspx.

74 "Does it matter how life is formed?": AL conversation with Ruth
    Faden, August 14, 2019.

75 "morality, value, and dignity": Yos Hut Khemacaro, email sent to
    AL, August 15, 2019.

75 "We're at or very near a cusp": Richard Hayes, email sent to AL,
    August 10, 2019.

76 "There is a serious concern": Paul Berg et al., "Potential Biohaz-
    ards of Recombinant DNA Molecules," *Science*, July 26, 1974.

77 "The Venter Institute's research": Presidential Commission for
    the Study of Bioethical Issues, December 2010, https://bioethics
    archive.georgetown.edu/pcsbi/synthetic-biology-report.html.

77 "I don't see that it makes any point": Richard Feynman, *The Plea-
    sure of Finding Things Out* (Cambridge, MA: Helix Books, 1999),
    p. 12.

78 "Although the company was not a business success": Szostak, No-
    bel autobiography.

ONE HUNDRED BILLION

84 The British philosopher Colin McGinn has argued: See, for example, his book *The Mysterious Flame* (New York: Basic Books, 1999).

SMILE

87 at the rate of ten trillion particles of light per second: I calculated the number of light particles (photons) reflected from the woman and entering the man's pupils in the following way: in diffuse daylight, the average light intensity is about 1.4 million ergs per second per square centimeter. Using the average energy of a visible light photon, two electron volts (1 erg = 6.24×10^{11} electron volts), this translates to 400 thousand trillion photons per square centimeter per second. Since a pupil is about 0.04 square centimeters in area in bright light, this gives 30 thousand trillion photons per second of *total* light entering the pupils. Now, assuming the woman has a body area of five square feet (a woman of modest proportions), at a distance of twenty feet she subtends about 0.002 of the hemisphere viewed by the man. The fraction of incoming light she reflects is about 20 percent. Taking this fraction of the total light (0.002 × 20% × 30 thousand trillion) gives the figure quoted.

87 Here it is gathered by one hundred million rod and cone cells: The structure of the eye, including the dimensions of the rod and cone cells, may be found in chapter 13 of *Gray's Anatomy*.

87 meeting a retinene molecule: Discussion of retinene molecules can be found in Allen Kropf and Ruth Hubbard, "Molecular Isomers in Vision," *Scientific American,* June 1967. The number of retinal molecules hit by photons per second quoted here is the total number, not just those reflected from the woman, and is calculated in the first note for this chapter.

88 First in the eye and then in the brain: For a discussion of transmission of visual information to the brain, the working of neu-

rons, the optic nerve, and the visual cortex, see David H. Hubel and Thorsten N. Wiesel, "Brain Mechanisms of Vision," *Scientific American,* September 1979.

89 The sound makes the trip: The speed of sound in air under normal conditions is 1,100 feet per second.

89 The eardrum, an oval membrane: The anatomy of the ear can be found in chapter 13 of *Gray's Anatomy.*

THE ANATOMY OF ATTENTION

92 "we were satisfied to know which areas": I interviewed Robert Desimone in his office at MIT on September 17, 2014.

92 Desimone and his colleague Daniel Baldauf reported: "Neural Mechanisms of Object-Based Attention," *Science* 344, no. 6182 (April 2014): 424–27.

IMMORTALITY

101 The neuroscientist Antonio Damasio: See, for example, *The Feeling of What Happens: Body and Emotion in the Making of Consciousness* (New York: Harcourt, 1999).

THE GHOST HOUSE OF MY CHILDHOOD

108 "Dream delivers us to dream": Ralph Waldo Emerson, "Experience," in *Essays: Second Series* (1844).

IN DEFENSE OF DISORDER

110 The Buddhist monks from Namgyal monastery: See https://www.youtube.com/watch?v=dORgAH1qDF8.

111 "However we analyze the difference": E. H. Gombrich, *The Sense of Order,* 2nd ed. (London: Phaidon, 1984), p. 9.

112 "Any body wholly or partially immersed in a fluid": Archimedes's law of floating bodies: can be found in *The Works of Archimedes,* edited by T. L. Heath (New York: Dover, 2002), p. 253. See also https://en.wikipedia.org/wiki/On_Floating_Bodies. Also

https://www.stmarys-ca.edu/sites/default/files/attachments
/files/On_Floating_Bodies.pdf.

113 resembling a satyr more than a man: For physical descriptions of
Socrates, see Plato's *Theaetetus* 143e, and *Symposium* 215a–c, 216c–d,
221d–e; Xenophon's *Symposium* 4.19, 5.5–7; and Aristophanes's
Clouds 362.

113 "He who, having no touch": *The Dialogues of Plato*, vol. 7, trans.
Benjamin Jowett (Chicago: Britannica Great Books, 1987), p. 124.

113 "For 15 days, I strove to prove": Henri Poincaré, *The Foundations
of Science*, trans. George Bruce Halsted (New York: Science Press,
1913), p. 387.

114 He was born in 1822 in Pomerania: Much biographical information on Clausius can be found in his entry in the *Complete Diction-
ary of Scientific Biography*, 26 vols. (New York: Charles Scribner's
Sons). See also the following note.

114 "A chief characteristic was his sincerity and fidelity": "Obituary
Notices" of the *Proceedings of the Royal Society of London*, vol. 48.

114 Clausius's great paper on disorder: An English translation of
Clausius's 1850 paper, "On the Moving Source of Heat," can be
found in *A Source Book in Physics*, trans. William Francis Magie
(Cambridge, MA: Harvard University Press, 1969), pp. 228–36.
The original title of the work, including the word *Wärme*, is given
here.

116 Rather, there is a slight asymmetry: I am referring here to something called "CP violation in physics." You can read about this,
and how it leads to the imbalance in particles and antiparticles
in many places: https://en.wikipedia.org/wiki/CP_violation.

119 "the wanderlust gene": See, for example, Gavin Haines, "The
'Wanderlust Gene'—Is It Real and Do You Have It?" *The Tele-
graph*, August 3, 2017.

119 "We have evidence to suggest": See ibid. One of Ebstein's original
research papers is R. P. Ebstein et al., "Dopamine D4 Receptor
(D4DR) Exon III Polymorphism Associated with the Human

Personality Trait of Novelty Seeking," *Nature Genetics* 12 (1996): 78–80, doi: 10.1038/ng0196-78.

120 I am listening to Anton Bruckner's Ninth Symphony: You can listen to Bruckner's symphony in many places. For example: https://www.youtube.com/watch?v=M4IUfuNV12c.

MIRACLES

123 A 2013 Harris poll found: https://theharrispoll.com/new-york-n-y-december-16-2013-a-new-harris-poll-finds-that-while-a-strong-majority-74-of-u-s-adults-do-believe-in-god-this-belief-is-in-decline-when-compared-to-previous-years-as-just-over/.

128 "The disharmony in our relationship to the Earth": Al Gore, *Earth in the Balance* (Boston: Houghton Mifflin, 1992), p. 223.

129 "I believe that our physical universe": Owen Gingerich to AL, July 7, 2011.

130 "the belief that there is an unseen order": William James, *Varieties of Religious Experience* (1902). See Lecture 2, BiblioBazaar edition (2007), p. 60.

OUR LONELY HOME IN NATURE

137 The 2014 report by the United Nations Intergovernmental Panel on Climate Change: https://www.ipcc.ch/report/ar5/wg2/.

IS LIFE SPECIAL?

139 "Our team is thrilled to be a part": "Kepler Mission Rockets to Space in Search of Other Earths," March 6, 2009, https://science.nasa.gov/science-news/science-at-nasa/2009/06mar_keplerlaunch.

140 I estimate that the fraction: My calculation of the fraction of matter in living form: Experts have estimated the biomass of the Earth at about 2×10^{18} grams. The typical star with a habitable planet is about 0.2 the mass of the Sun. That gives a fraction of biomass in a typical habitable solar system of about 2.5×10^{-15}. The fraction of mass in the universe in solar systems (visible form) is about 0.05.

(The rest is dark matter and dark energy.) One-tenth of all stars have habitable planets. Combining these numbers gives a fraction of about 10^{-18}. To compare to the Gobi Desert, I take the Gobi Desert to be about 500,000 square miles in area, and a typical grain of sand to have an area of about 2×10^{-3} square centimeters.

142 Yet, polls of the American public show: See, for example, this 2015 study by Pew Research Center: https://www.pewresearch .org/fact-tank/2015/11/10/most-americans-believe-in-heaven -and-hell/.

COSMIC BIOCENTRISM

149 "It is impossible to calculate in detail": Freeman Dyson, "Time Without End: Physics and Biology in an Open Universe," Review of Modern Physics 51, no. 3 (July 1979): 447.

150 he quotes another great theoretical physicist: Steven Weinberg, *The First Three Minutes* (New York: Basic Books, 1977).

150 "This is often the way it is in physics": Ibid., p. 131.

150 His older sister Alice remembers: Ann Finkbeiner, "Freeman Dyson Turns 90," *The Last Word on Nothing* (blog), October 7, 2013, https://www.lastwordonnothing.com/2013/10/07/freeman -dyson-turns-90/.

157 If you swipe the mountain with a piece of silk: "Kalpa," *Chinese Buddhist Encyclopedia,* http://www.chinabuddhismencyclopedia .com/en/index.php/Kalpa.

158 "It is in the highest degree unlikely": Lucretius, *De Rerum Natura* (Cambridge, MA: Harvard University Press, 1982), book II, lines 1055–57 and 1074–78.

161 "the reverence for life": See the 1952 Nobel Peace Prize Award Presentation, https://www.nobelprize.org/prizes/peace/1952/ ceremony-speech/, and by Albert Schweitzer, *Out of My Life and Thought,* trans. C. T. Campion (New York: Holt, Reinhart, and Winston, 1949), p. 157.

161 "I am life that wills to live": Ibid.

THE MAN WHO KNOWS INFINITY

164 If a weary traveler arrives: The concept of the infinite hotel is called "Hilbert's Hotel" and was first introduced by the German mathematician David Hilbert to convey some of the non-intuitive properties of infinity.

165 The first recorded conception of infinity: For more on early Greek ideas on infinity, see Elizabeth Brient, *The Immanence of the Infinite: Hans Blumenberg and the Threshold to Modernity* (Washington, DC: Catholic University of America Press, 2002).

165 the Chinese employed the word *wuji:* For more on the Chinese conception of infinity, see Jiang Yi, "The Concept of Infinity and Chinese Thought," *Journal of Chinese Philosophy* 35, no. 4 (December 2008): 561–70.

165 Aristotle argued that infinity: For Aristotle's discussion of potential versus actual infinity, see his *Physics*, book 3, chapter 6.

166 "God is everywhere, and in all things": *Summa Theologiae*, I.7.1, translated by Fathers of the English Dominican Province (Benziger Brothers, 1947).

166 "It's kind of like mathematics": Denise Chow, "No End in Sight: Debating the Existence of Infinity," *Live Science*, June 3, 2013, https://www.livescience.com/37077-infinity-existence-debate .html.

167 In the early 1980s, Linde proposed: Linde's first papers on inflation are A. D. Linde, "A New Inflationary Universe Scenario: A Possible Solution of the Horizon, Flatness, Homogeneity, Isotropy and Primordial Monopole Problems," *Physics. Letters B* 108, 389 (1982); "Chaotic Inflation," *Physics Letters B* 129, 177 (1983); "Eternally Existing Self-reproducing Chaotic Inflationary Universe," *Physics Letters B* 175, 395 (1986).

167 a revision of MIT physicist Alan Guth's 1981 theory: Guth's original paper on the inflation theory is A. Guth, "Inflationary Universes: A Possible Solution to the Horizon and Flatness

Problems," *Physical Review D* 23:347 (1981). Note: the "horizon" problem is what I later discuss as the problem of the uniformity of the cosmic background radiation.

169 "I easily understood what Guth was trying to do": Interview with Linde, October 22, 1987, Cambridge, MA, printed in Alan Lightman and Roberta Brawer, *Origins: The Lives and Worlds of Modern Cosmologists* (Cambridge, MA: Harvard University Press, 1990), pp. 486–87.

169 "Do you think dinosaurs truly existed?": My recent interview with Linde was at his home in Stanford, California, on July 10, 2019. All quotes from Linde at this time are taken from this interview.

173 "At the first glance, this theory seemed to be too exotic": Autobiography of Andrei Linde for the Kavli Foundation, 2014, http://kavliprize.org/sites/default/files/Andrei%20Linde%20autobiography.pdf.

177 the Babylonian Map of the World: See https://www.ancient.eu/image/526/babylonian-map-of-the-world/. See also https://en.wikipedia.org/wiki/Babylonian_Map_of_the_World.

178 "Every universe is covered by seven layers": Bhagavata Purana 6.16.37. See https://prabhupadabooks.com/sb/6/16/37.

179 The resulting video, made by Stanford University: https://www.youtube.com/watch?v=ZlfIVEy_YOA.

180 In a *New York Times* article: Dennis Overbye, "Space Ripples Reveal Big Bang Smoking Gun," *New York Times,* March 17, 2014.

181 One story he enjoys telling: AL interview with Linde, October 22, 1987. This story does not appear in the abridged version of the interview published in Lightman and Brawer, *Origins,* but does appear in the full interview, which can be found at https://www.aip.org/history-programs/niels-bohr-library/oral-histories/34321.

THE ACCIDENTAL UNIVERSE
The World You Thought You Knew

With passion and curiosity, Alan Lightman explores the emotional and philosophical questions raised by recent discoveries in science. He looks at the dialogue between science and religion; the conflict between our human desire for permanence and the impermanence of nature; the possibility that our universe is simply an accident; the manner in which modern technology has separated us from direct experience of the world; and our resistance to the view that our bodies and minds can be explained by scientific logic and laws. Behind all of these considerations is the suggestion—at once haunting and exhilarating—that what we see and understand of the world is only a tiny piece of the extraordinary, perhaps unfathomable whole.

Science/Philosophy

ALSO AVAILABLE

The Diagnosis
The Discoveries
Einstein's Dreams
Ghost
Mr G
Reunion
A Sense of the Mysterious